SpringerBriefs in History of Science
and Technology

For further volumes:
http://www.springer.com/series/10085

Donna J. Drucker

The Machines of Sex Research

Technology and the Politics of Identity, 1945–1985

 Springer

Donna J. Drucker
Technische Universität Darmstadt
Darmstadt
Germany

ISSN 2211-4564   ISSN 2211-4572 (electronic)
ISBN 978-94-007-7063-8   ISBN 978-94-007-7064-5 (eBook)
DOI 10.1007/978-94-007-7064-5
Springer Dordrecht Heidelberg New York London

Library of Congress Control Number: 2013941345

Printed on acid-free paper

Springer is part of Springer Science+Business Media (www.springer.com)

*To the memory of Peter A. Kraemer
(1972–2010)*

Acknowledgments

The inspiration for this book came from a December 2010 visit to a sex research laboratory at the Kinsey Institute for Sex, Gender, and Reproduction Inc., Bloomington, Indiana. The psychophysiologist Erick Janssen showed me the room. It was in a quiet corner of the Institute, and included a soft armchair, a computer LCD monitor with keyboard, a set of drawers along one side that contained measurement devices, and an intercom system behind the chair where a researcher could communicate with a participant without the participant seeing them. There were curtains over the windows. My eyes were drawn to some of the smaller machines on top of the drawers, and I wanted to look more closely at them. But Dr. Janssen drew my attention to a space heater set next to the armchair. "This," he proclaimed, "is the most important piece of equipment in this laboratory." In that moment, I had a glimpse of the myriad ways that investigators and subjects interact with machines and objects in laboratories to make human sex research possible, and I knew that I wanted to understand the story of how present-day sexual physiology research came to be.

Many thanks are due to people and institutions for their assistance along the way: For making publication with Springer possible, Charles T. Wolfe, Lucy Fleet, and the anonymous reviewer; for funding support and the time to write, the Topologie der Technik Graduiertenkolleg at Technische Universität Darmstadt, Germany; and for insightful discussions of the histories of technology, gender, and sexuality, Tobias Boll, Jessica Martucci, Arwen Mohun, Suzanne Moon, Eric Shatzberg, Lee Vinsel, and the professors and graduate students in the Graduiertenkolleg, especially Mikael Hård. Special thanks to Julia R. Heiman for taking the time for an interview; to Tommy Dickinson and Hallie Lieberman for sharing their unpublished work with me; and to Mark A. Price for reading and commenting on the whole manuscript. I am grateful to friends John Philipp Baesler, Kathleen A. Costello, Rachel E. Feder, Stefan Glatzl, Robin C. Henry, Anne Lucke, and Jennifer Stinson, and to my family, especially my parents, for their care and love. I dedicate this book to the memory of Peter A. Kraemer.

Archival material from the Virginia Johnson Masters Collection is reprinted courtesy of the Kinsey Institute for Research in Sex, Gender, and Reproduction Inc., Bloomington, Indiana, USA.

Archival material from the Women's Liberation Movement: An Online Archival Collection is reprinted courtesy of the David M. Rubenstein Rare Book and Manuscript Library, Duke University, Durham, North Carolina, USA.

There are instances where I have been unable to trace or contact the copyright holders. If notified, the publisher will be pleased to rectify any errors or omissions at the earliest opportunity.

Contents

Chapter 1
The Machines of Sex Research

Abstract The histories of gender, sexuality, and technology in the twentieth century link together in the context of human sex research. The development and use of specific machines in sex research began in 1896 and continues through the present. A relationship exists between the type of machines scientists invented, the historical context in which scientists operated, and the type of research that they published. This chapter describes the theoretical background for the book as a whole, overviews its contents chapter by chapter, and outlines the history of machines used in sex research from 1896 through the mid-1950s. The theoretical background combines spatial, gender, sexuality, and technology theories. The overview describes the following chapters: Chap. 2, which focuses on the development of the sex research laboratory, the penile strain gauge, and the rise and fall of aversion therapy; Chap. 3, which centers on the work of William Masters and Virginia Johnson on singles and couples using a penis-camera, published as *Human Sexual Response* (1966); Chap. 4, which describes the development of machines specifically for sex research on women, such as the vaginal photoplethysmograph and the labial thermistor; and the conclusion, which brings the narrative up to the present. The history of machine-based sex research from 1896 through the mid-1950s includes discussion of Martin Mendelsohn, John B. Watson, Alfred Kinsey, and Wilhelm Reich.

Keywords Sexology · Sex research · Technology · Machines · Alfred Kinsey · William Masters

Human sexuality and technology have an intertwined and complex history from the late nineteenth century to the present. Women in the late Victorian era used machines to provide themselves orgasmic relief from "hysteria" (Maines 1999; Lieberman 2011). Inventors created, and continue to create, machines to help men with erectile dysfunction and premature ejaculation or to assist women with vaginal blood flow and lubrication (McLaren 2007; Lieberman 2011). Surgeons use certain tools to shape or to "fix" genitals into idealized contours and sizes. Countless machines are available through sex-toy shops and companies to assist

D. J. Drucker, *The Machines of Sex Research*,
SpringerBriefs in History of Science and Technology,
DOI: 10.1007/978-94-007-7064-5_1, © The Author(s) 2014

individuals, couples, and groups increase their chances of sexual pleasure. Machines have also, however, become essential components of scientific human sex research dating from the end of World War II. The use of specifically genital measurements, such as penile strain gauges and vaginal photoplethysmographs, is part of a broader historical shift in the use of machines in sex research from the mid-1940s onward, including punched-card machines, respirators, electrocardiograms, penis-cameras, film cameras and projectors, tape recorders, televisions, slide projectors, audiotapes, portable space heaters, and VCRs/DVD players.

As researchers integrated machines into sex research across the Western world in the postwar era, they incorporated machines into the broad medical programs, social structures, and political mechanisms that policed and enforced reproductive heterosexuality from the 1890s through the late 1960s. They also used machines to form and to impose ideals of sexual and gender normativity on the interior and exterior spaces of male and female bodies. Researchers assayed a variety of devices, some that worked well, and many others that did not, and by the end of the 1970s had for the most part developed a standardized set of practices for basic sexual physiology investigations. Though as some researchers became more attentive to the diversity and complexity of the relationships between types of arousal in the body and mind, and as global gay and lesbian rights activists advocated for the right to public recognition of their identities and personhood from the late 1960s onward, researchers slowly became more focused on women's sexual responses or on comparing the responses of men and women. Thus as newer technologies of sex research became more holistic in light of social movements for gay and lesbian civil rights, researchers and activists used those technologies to re-envision sexual and gendered bodies as sites of non-reproductive orgasmic pleasure.

Second-wave feminism, gay and lesbian rights movements, and civil and democratic rights movements broadly provided the context for the shift in the focus in sex research from punitive measures to investigations of basic physiology and numerous other topics. Radical feminism galvanized the American feminist movement in the late 1960s and early 1970s, cementing key concepts such as "the myth of the vaginal orgasm" and "the personal is political" into popular discourse and consciousness (Faludi 2013). Women worldwide made numerous political, economic, and cultural gains from the late 1960s through the mid-1980s, including the passage of Title IX in the U.S. requiring gender equality in college sports, contraception becoming free in the United Kingdom, formal women's organizations being founded in Turkey and Israel, Swiss women gaining the right to vote nationally, and an Equal Employment Opportunity Law in Japan passing in 1986. However, a major aim of the second-wave feminist movement in the United States, the passage of the Equal Rights Amendment (ERA) to the Constitution, was defeated in 1982, three states short of ratification (Mansbridge 1986; Berry 1988). The beginning of the "feminist sex wars" at a Barnard College conference on sexuality muddled the coherence of second-wave feminism, dividing it into numerous factions, such as pro- and anti-pornography, pro- and anti-prostitution, and libertarian and radical (Vance 1984; Gerhard 2001). Nonetheless, second-wave feminism broadly, together with homosexual liberation movements, informed the thinking of sex researchers, inspiring them to create machines that supported their studies for physiological and psychological research into women's sexualities across spectrums of identity, behavior, and experiences.

Changes in the legal status of homosexuality affected researchers' outlooks and perspectives on their work as well. Worldwide laws against homosexual behavior were overturned beginning in the early 1970s with the rise of gay and lesbian liberation movements. Britain, Wales, and Canada decriminalized homosexual acts between men in 1967 (Blasius and Phelan 1997). Homosexuality was decriminalized in Austria, Costa Rica, and Finland in 1971. That was the same year that homosexual rights advocacy movements in Australia (Society Five) and in the United Kingdom (Gay Liberation Front, or GLF) were founded (Power 1995; Cook 2007). In a widely reported professional statement, the American Psychological Association (APA) removed homosexuality as a disorder from the *Diagnostic and Statistical Manual* (DSM-II) in 1973 (This American Life 2002). Individual municipalities and states throughout the United States began to repeal anti-sodomy legislation and to pass laws forbidding discrimination based on sexual orientation. There was conservative pushback against those laws, most famously by Anita Bryant's successful campaign against an antidiscrimination ordinance in Dade County, Florida in 1978, and by an unsuccessful attempt to pass a statewide ban on homosexual teachers (Proposition 6, aka the Briggs Initiative) in California in the same year (Shilts 1988; Marcus 2002; Milk 2012). Sodomy was not declared illegal in the U.S. nationwide until the *Lawrence v. Texas* Supreme Court decision in 2003 (Carpenter 2012).

Sex research is one of the intellectual elements that shapes political and cultural concepts of sexuality and gender. The machines that sex researchers created and used concretized their perceptions of gender and sexuality in the bodies they studied. The numerous kinds of machines developed specifically for sex research over the past century show the imagination of sex researchers and their strong interest in discovering some facts or truth about human sexuality. Sometimes that interest led to harm to their subjects, and sometimes it led to broader and better understanding of human sexuality. Electrical and battery-powered machines are now essential parts of human sex research.

Chapter Overview

Other authors have looked at the use of machines in sex research, but none have clearly connected the development of machines to political and cultural developments. Such authors include Mary Roach, who used some historical material in her journalistic perspective on sex research, *Bonk* (Roach 2008). Hoag Levins, in *American Sex Machines,* described multiple patents from the 1840s through the 1980s in the U.S. Patent and Trademark Office for sex-related, including sex-preventative, clothing and devices (Levins 1996). Timothy Archibald's *Sex Machines: Photographs and Interviews,* contained a collection of photographs of homemade sex machines and their creators, all white men from the rural United States (Archibald 2005). Some histories of sexuality briefly mention that machines were involved in sex research, but do not take the machines themselves as an object of study (D'Emilio and Freedman 1988; McLaren 1999; Garton 2004). A focus on innovation in machines used for sex research shows how developments in machine design, cultural forces, and researchers' perspectives on their work are linked. This book is primarily focused on the United States and United Kingdom, though researchers in other countries, namely Germany, Canada, the former Czechoslovakia, and Israel are also part of the narrative. The focus on the United States and United Kingdom is largely because those countries are where most of the sex research devices that have had a lasting impact on present-day sex research were developed. The remainder of this introductory chapter outlines the theoretical background of this book, provides a short history of machines used in sex research from 1896 through the 1950s, and suggests reasons why machine-based sex research was so infrequent before the 1960s.

The second chapter describes the creation of mechanical devices for men from the 1950s through the early 1970s: some to detect and to measure men's erections, and others that researchers used to police and to discipline men considered sexually deviant. Researchers in Germany in 1936 developed a penile strain gauge made of metal, and one in Czechoslovakia made a related device, the vacuum pump, to measure circumferential and volume changes in men's erections. Physicians interested in sexual deviance, whether it was fetishism, homosexuality, cross-dressing, or pedophilia, paired metal or mercury-in-rubber strain gauges with slides, audio recordings, and films, as well as with electroshock machines, in an attempt to use electric power along with the power of the state to constrain or to punish men into reproductive heterosexuality. Physicians on both sides of the Cold War had the moral authority to try to make the body politic into the "straight state," and the treatments continued through the early 1970s despite their poor to nonexistent results (Canaday 2009). Treatments gradually ceased after the 1973 APA decision to remove homosexuality from its list of psychological disorders, and after many homosexual acts in Britain were legalized in 1967 (This American Life 2002; Cook 2007; Cohen 2013). Nonetheless, the harsh treatments had a long-term impact on the patients, and on many of the doctors and nurses who administered them (Dickinson 2012).

The third chapter focuses on William H. Masters and Virginia E. Johnson, two St. Louis, Missouri–based sex researchers active from the 1950s through the mid-1990s who adopted multiple technologies into their laboratory studies. Masters, an obstetrician and gynecologist specializing in fertility, turned to sex research when he realized how little physicians understood about the relationship between sexual behavior and fertility. Masters and his assistant (later wife) Virginia Johnson developed research protocols involving respiratory, blood pressure, and other machines to study the physiology of men's and women's sexual behavior, alone and in pairs. They and Washington University technicians invented a unique technology known as the penis-camera, an electric film camera inside a Plexiglas dildo that filmed women's vaginas and cervixes as women masturbated. From the penis-camera's data and those from the other machines, Masters and Johnson developed a four-stage sexual response cycle for men and women, which they argued made orgasmic and sexual equality a possibility for heterosexual married couples. Feminist thinkers read Masters and Johnson's *Human Sexual Response* and were inspired to integrate sexual fulfillment and satisfaction for women of all sexual orientations into second-wave feminist writing (Masters and Johnson 1966; Gerhard 2001). Also, Masters and Johnson's work inspired psychologists to investigate and to expand further on the results in *Human Sexual Response*.

The fourth chapter outlines the development of the vaginal photoplethysmograph and related tools specifically made for measuring women's sexual physiology. The vaginal photoplethysmograph was developed independently in Israel in 1965 and in the United States in 1974, as ideas of second-wave feminism rose to prominence worldwide (Bercovici and Palti 1967; Palti and Bercovici 1967; Geer et al. 1974). The vaginal photoplethysmograph, also known as a vaginal photometer, measures vaginal blood volume and vaginal pulse amplitude using a photocell embedded in a small glass tube. Devices developed later in the 1970s and early 1980s also included a labial clip that measured labial temperature using a thermistor (Henson et al. 1977). Some researchers tried one-time experiments using uterine balloons or other painful devices that were unwieldy, expensive, or caused subjects much physical distress. Technologies for women in the present, though standardized, remain unsatisfying for many current researchers. Devices that measure the full range of women's physiological arousal are still in the future.

The book's conclusion brings the story of sex research machines to the present, as many questions raised in the earliest days of sex research are still a concern for scientists. Machines used in sex research are fully integrated not only into academic research, but also into popular, consumer, and legal culture. The vacuum pump is still a part of the American legal system, as local jurisdictions use it to measure the erectile responses of convicted sex offenders. Machines that were previously used against a medical subjects' will are now a common part of amateur and professional pornography, and are also now widely available from erotic product purveyors. As machines become more and more a part of people's sex lives online and offline, sex researchers keep seeking and testing professional devices that will help them better understand the mechanisms of arousal and the relationship of mind and body.

Theoretical Background

Studying the machines of sex research draws attention to the ways that those machines delineate, categorize, expand knowledge of, and police bodily spaces both public and private, inside and out. As Henri Lefebvre has argued, all social processes have a fundamental spatiality, and the human body itself is a factor in the production of space (Lefebvre 1991). The machines of sex research affect scientific perceptions of the human body; the laboratory and office spaces in which researchers use them; the physical and emotional spaces where human relationships take place; metaphorical and material conceptions of the "closet" for homosexual and queer persons; and the public spaces of neighborhoods, workplaces, and landscapes in which people interact with each other. Machines were tools of more hostile spaces, such as laboratories in which people received aversion therapy; they were tools that reinforced heterosexuality in the laboratory; they perpetuated images of the interior of bodies as sexual spaces; and they provided data on the human body that changed how individuals interacted with and used public and private space in their everyday lives. Machines could play varying roles and could support various research focuses and agendas. "By borrowing well-established knowledge, and by incorporating it in pieces of furniture or in routine operational sequences, the laboratory can harness the enormous power of tens of other fields for its own purposes" (Latour and Woolgar 1986, p. 68).

Using machines for sex research raises questions about how scientific understandings of the interior and exterior of bodies affect what it means to have a sexual body, the use and meaning of pain in human research, and the degree to which subjectivity is possible for the sex research participant. As Elizabeth Grosz writes, "knowledges require the interaction of power and bodies; correlatively, power requires knowledges of bodies and behaviors in order to remain effective....Bodies are thus essential to accounts of power and critiques of knowledge" (Grosz 1995, p. 32). The bodies of subjects were and are essential to the production of sex research, as they stand in for the bodies of wider populations whom researchers intend to investigate, fix, or heal. As this book shows, bodies often do not behave how researchers expect them to, and they can confirm or can challenge scientists' expectations and perceptions of bodies and sexualities, along with their own sense of control and knowledge.

Further, this book addresses the broad theme of how the availability and development of instruments has affected sexual scientists' thinking about what kinds of research—not to mention what kinds of results—are possible. Peter Galison issued a call for research on instruments and their meanings in 1987 when he wrote, "Too little is known about the ways in which scientific beliefs re-create themselves in instrument design. The history of instruments that we need must be an archaeology that uses the material culture of science to unearth buried theoretical assumptions and experimental practices" (Galison 1987, p. 252). For Galison, machines imported assumptions that scientists consciously or unconsciously built into them, and the historian of science must uncover these "technological assumptions" to show that "machines are not neutral" (ibid.). Or as John Staudenmaier put it, "Technological artifacts exist as crystalized moments of past human vision, each one a little 'master narrative' seeking to enforce its perspective, each one buffeted by the swirl of passion, contention, celebration, grief, and violence that make up the human condition" (Staudenmaier 1994, p. 273).[1]

---

[1] Michel Foucault framed his ideas regarding how to approach apparatus historically slightly differently: "What I'm trying to pick out with this term [apparatus] is, firstly, a thoroughly

Focusing attention on the creation, testing, and changes in devices developed and used for sex research shows their historical context and contingency. Although some devices never clearly emerge onto the research stage or vanish almost entirely—such as John B. Watson's mysterious instruments that were hidden away for decades—other devices, such as the penile strain gauge, went through cycles of apolitical and political uses depending on medical and state interests and sociocultural contexts. In other words, machines used in human sex research were never neutral. The meanings that clinicians, nurses, patients, research volunteers, and reading audiences ascribe to them have fluctuated dramatically over time.

The context of the development, use, reception, and modification of an apparatus itself also reveals much about the individuals and social groups that had some interaction with it. Delineating the history of machines used in human sex research also shows how unexpected the uses of an object can be after its initial introduction into scientific literature. Technological development of artifacts in sex research, rather than being a simple line drawing from object A to object B to object C, illustrates Wiebe E. Bijker, Thomas P. Hughes, and Trevor J. Pinch's point that technological development is most accurately modeled "as a non-determined, multidirectional flux that involves constant negotiation and renegotiation among and between groups shaping the technology" (Bijker et al. 1987, p. 13). Using a multidirectional model makes it more "possible to ask why some of the variants 'die,' whereas others 'survive.' One can expect to bring out more clearly the interpretative flexibility of technological artifacts" (Pinch and Bijker 1987, p. 29). The multidimensional nature of sex research objects—how different actors perceive them, how their use changes through time, why some have stayed fixtures in sexology and others have disappeared—appears more clearly when their development is contextualized historically.

The theoretical framework in terms of sexuality and gender for this book derives from Judith Butler, Eve Kosofsky Sedgwick, and Michel Foucault (Butler 1990; Sedgwick 1990; Foucault 1977, 1980, 1985). If, as Foucault famously suggested, "the homosexual became a species" in the late nineteenth century, some sex researchers from the 1940s through the 1970s used their machines to punish and to discipline bodies then taxonomically marked as queer (Foucault 1977, p. 43). However, as homosexual rights movements gained momentum worldwide, other sex researchers from the mid-1970s onward used their machines to disrupt standardized narratives of heterosexual performativity and to support non-binary sexed and gendered identities, desires, and behaviors. As Michael P. Brown argued, the

---

(Footnote 1 continued)
heterogenous ensemble consisting of discourses, institutions, architectural forms, regulatory decisions, laws, administrative measures, scientific statements, philosophical, moral and philanthropic propositions—in short the said as much as the unsaid. Such are the elements of the apparatus. The apparatus itself is the system of relations that can be established between these elements… The apparatus is thus always inscribed in a play of power, but it is also always linked to certain coordinates of knowledge which issue from it but, to an equal degree, condition it" (Foucault 1980, pp. 194–196).

closet is not just a metaphor for concealment of queer presence, but it also operates as a material and spatial practice on multiple scales (Brown 2000). While earlier sex researchers used their machines to keep the closet door firmly shut, later sex researchers used theirs to open interior and exterior bodily spaces outward into bedrooms, relationships, and public spaces to promote exploration and pleasure in sexed and gendered expressions of personhood. The study of machines in sex research thus creates "a new critical human geography" showing how at different times in history those machines helped researchers first police but then expand sexed and gendered forms of being (Soja 1989, p. 6).

The above-described bodies of literature provide a background to study the ebbs and flows in the machinery of sex research from the postwar era through the mid-1980s. Researchers who focused on men centered their efforts on producing and perfecting devices that would produce measurements, statistics, diagnoses, and treatment protocols that would turn deviant, queer, and impotent men into heterosexual fathers. Homosexuality, pedophilia, transsexuality, cross-dressing, and impotence blurred together into a broad category of social deviance that many sex researchers through the early 1970s thought they could fix with the additional electroshock machines used for aversion therapy. Sex researchers using machines on men alone between 1945 and 1980 viewed erections as a physiological event needing mechanical and medical mediation so that they would only occur in reproductive, heterosexual circumstances or in sleep. For those researchers, the blood vessels inside the penis and the penis itself were both sexualized spaces needing mechanical management, and men who participated in these research projects were subject to painful restraints around their genitals and nausea, vomiting, discomfort, and psychological distress as a result of aversion therapies. Those researchers viewed the interior and exteriors of their male-bodied subjects as dangerous and unruly, spaces needing control for proper functioning in Western society. Queer men in the gay rights movement of the 1970s and early 1980s would reject these techniques as interfering with their rights to dignified personhood accompanying their political and social gains.

However, the machinery of sex research for women developed not only during the homosexual rights movements, but also during the second global wave of feminism (D'Emilio and Freedman 1988). "In the long wake of the surveys of Alfred Kinsey, the studies of William Masters and Virginia Johnson, the sexual liberation movement and the rise of feminism, there has been a surge of scientific attention, paid by women, to illuminating the realm of women's desire" (Bergner 2009). While the lack of an accurate instrument initially hampered research on women, the development of the dildo-camera in the early 1960s and the vaginal photoplethysmograph in the U.S. 1974 directed attention toward comparing their arousal (in the form of vaginal blood flow and pulse amplitude) with men's or toward studying it on its own. For example, Masters and Johnson's Plexiglas penis-camera mapped the physiological changes in the interior of women's vulvas and disproved Sigmund Freud's theory of orgasmic transfer, showing that the clitoris was the primary organ for women's pleasure across their lifetimes (Masters and Johnson 1966; Maines 1999).

Machines used to study women or men and women together were less painful and restrictive and did not involve using punishment to discipline the subjects' bodies. The machines in that phase of sex research, such as cardiographs and blood pressure monitors recording laboratory subjects engaging in masturbation or intercourse, may have been awkward or embarrassing, but researchers used them discreetly and non-invasively for information gathering. As women worldwide advocated for sexual and reproductive freedom and for civil rights broadly, researchers further developed technologies for measuring women's arousal that aimed to create the basic body of knowledge about women's sexuality needed for applied research. Even though researchers like Johnson dismissed the applicability of her and Masters' findings for women's and lesbian rights movements, members of those movements nonetheless appropriated such research to support their rights to bodily autonomy, agency, and non-reproductive pleasure. Unlike the work done on men alone in the 1960s and 1970s, comparative or female-only research in the mid- to late 1970s was not intended to change the behavior of the subjects, but rather to understand it as it was—and to reclaim women's bodies as spaces of pleasure beyond reproduction alone.

Historical Background

The use of machines for research on sex from the 1890s through the mid-1950s tended to make researchers pariahs among their peers. Sex research was published in obscure journals if at all, or addressed sex only obliquely as part of another research interest. Sex research at the turn of the last century was psychologically oriented: prominent researchers such as Sigmund Freud, Carl Jung, Albert Moll, Richard von Krafft-Ebing, Havelock Ellis, and Magnus Hirschfeld developed their theories about human sexuality through oral patient interviews and treatments, through written surveys (especially for Hirschfeld) and through correspondence (especially for Ellis). If a person's sexual issues and concerns were rooted in the mind, it was of lesser concern for researchers to study a person's physical body. Other scholars have well documented the interactions between the most well-known sex researchers and their influences on each other (Gay 1988; Mancini 2010; Sigusch 2012; Sauerteig 2012; Oosterhuis 2000; Oosterhuis 2012).

The Brooklyn, New York–based doctor Robert Latou Dickinson and his assistant Lura Ella Beam published some limited investigations about Dickinson's work as an obstetrician and gynecologist using non-mechanical tools such as speculums and glass dildoes in the 1930s (Dickinson and Beam 1931; Dickinson and Beam 1934). Dickinson and Beam described an experiment in which Dickinson inserted a clear glass dildo into a female patient's vagina and observed the changes within as she masturbated to orgasm (Dickinson and Beam 1931, p. 93; Bullough 1994). However, until the publication of Alfred Kinsey's *Sexual Behavior in the Human Female* and *Human Sexual Response,* basic physiological research on women was virtually nonexistent (Kinsey et al. 1953; Masters and

Johnson 1966). At the same time, the availability of electrical machines for physiological investigations intrigued physicians who sought to make their own discoveries about the human body, and some of those physicians wanted to connect processes of the body with those of the mind. In the late nineteenth century, inventors built and tested all kinds of machines that could measure bodily processes with varying degrees of accuracy. Some of the most popular machines (which were also used for animal research) included a sphygmograph, which measured pulse rate; a plethysmograph, which measured change in limb volume using an enclosed cylinder; and the kymograph, a drum with a roll of smoked paper hooked to a stylus, that measured changes in blood pressure or respiration rate as the drum rotated. Physicians used those machines to measure changes in still or moving bodies to better understand their physiology, and to diagnose and to treat internal ailments. While inventors designed machines measuring different aspects of human and animal physiology throughout the nineteenth century, they only began publicly to report their use of those machines for studying human sexual physiology at century's end.

One of the first—if not the first—use of machines to study human sexual behavior appeared in the periodical *Deutsch Medizinische Wochenschrift (German Physician's Weekly)*. The periodical reprinted in six parts a lecture by the Berlin physician Martin Mendelsohn to an internal medicine interest group on the physiological effects of bicycle riding on men and women (Mendelsohn 1896a, b, c, d, e, f). Mendelsohn used a pneumograph, first developed by Étienne Jules Marey in 1876, to measure a bicycle rider's respiratory rate using a band wrapped around the subject's chest (Max Planck Institute for the History of Science 2008–2010). He also used a sphygmograph, likewise a Marey creation, which wrapped around a subject's wrist and forearm to measure a subject's pulse rate (ibid.). Both of those machines could be hooked up to a kymograph so that fluctuations in breathing and pulse rate could be recorded continuously for up to ten minutes. Mendelsohn was similarly interested in how bicycle riding's physiological effects interacted with a person's sexual behavior. He briefly described the case of a man who thought—albeit incorrectly—that strenuous bicycle riding had cured his gonorrhea (Mendelsohn 1896d). He noted that some male and female bicycle riders enjoyed their daily rides as masturbatory opportunities, as regular movement of the lower limbs caused friction of the penis or clitoris against the bicycle saddle. Sometimes men needed to stop bicycle riding because of the embarrassingly strong erections that they developed (ibid.). Mendelsohn also did some of his own testing of the relationship between the physiology of bicycle riding and sexual behavior. The exact set-up of his experiments was unclear, but he was able to produce three tables of data as a result: one with one pulse curve of a man who had sexual intercourse after a strenuous bicycle ride; one with three pulse curves showing a woman's pulse before, during, and after intercourse; and one with four pulse curves comparing one woman's and one man's pulse curves after the second of five sexual episodes within a short time period (Mendelsohn 1896e). As the subjects had sphygmographs around their wrists that were then attached to a kymograph, they did not have full freedom of movement, but their pulse rates nonetheless reached 150 beats per minute. Mendelsohn apparently did not repeat his sexual experiments, though he published further texts on medical apparatuses (Mendelsohn 1901). His results remained largely buried in the reams of medical literature on other aspects of human physiology at the turn of the last century, until Alfred Kinsey and his associates found and cited them in *Sexual Behavior in the Human Female* (Kinsey et al. 1953).

From Mendelsohn forward, mentions of technological sex research appeared sporadically in German and English medical literature. John B. Watson, one of the founders of behaviorism in psychology, lamented the lack of connection between psychological and physiological research needed to understand human emotions in his 1913 article that first outlined the basic principles of the field, "Image and Affection in Behavior" (Watson 1913). He pointed out that the machines currently in use for physiological research were not helping physicians answer the questions that he and others had about sexual physiology: "I have worked for years upon the expressive methods and no one will admit their failure in the past more readily than I. My present feeling is that we have taken our plethysmograms from the wrong organs. Whether there are too many technical difficulties in the ways of the objective registration of the many delicate changes in the sex organs remains for the future to decide" (ibid., p. 427). Watson believed that physicians were too limited in their abilities to measure sexual physiology, and so he conducted informal machine-based research behind closed doors in his Johns Hopkins University Laboratory, probably with a female graduate student, Rosalie Rayner. His wife found out and divorced him; he was fired from his position at the university; he married Rayner; and he spent the rest of his working life as a traveling salesman (Roach 2008). A picture of the instruments that Watson supposedly used with Rayner was printed in a 1981 article about Watson: they included a speculum, a tube-like instrument that could have been hooked to a kymograph, and two other devices with obscure functions and purposes (Magoun 1981).[2] As Watson's first wife destroyed his research papers, it is unclear what if anything he found in his laboratory sessions. What is clear is that conducting sex research for its own sake in the 1910s, without any other purpose or focus in mind, could be grounds for dismissal from academe.

Further scientific research through the first half of the twentieth century that involved measuring physical aspects of human sexuality tended to be buried in broader physiological studies. One example is of John C. Scott, a psychologist at the University of Pennsylvania, who studied the change in blood pressure in one

---

[2] Watson alluded to his findings in his use of machines for private sex research in a footnote to his 1913 article: "The whole area involved in sex function embraces a much wider zone than that of the sex organs proper. The erogenous areas are in infancy widely distributed throughout the body surfaces. Only gradually does the sex organ come to be looked upon as the focus of sex experience. Even in the case of most adults certain of these primitive zones remain functional, as, for example, the nipples, etc. The receptors lying in such areas are stimulated by the reflex motor processes initiated by the primary stimulus (i.e., the object under observation)" (Watson 1913, p. 427).

hundred men between the ages of nineteen and thirty-one as they watched three short films, each one depicting a situation designed to provoke feelings of sex, anger, or fear (Scott 1930). Scott also asked the men to rate the intensity of their feelings subjectively, on a five-point scale from very strong to very weak, so that he could compare their feelings with their blood pressure rates using a Tycos Recording Syphgmomameter. Scott found that the men's blood pressure rates were highest during the sex-related film, but speculated that they were probably too embarrassed to record the genuine high intensity of their subjective sexual arousal.

Another example is of Ernst P. Boas and Ernst F. Goldschmidt, who included sex-related data in their book *Studien mit dem Kardiotachometer über Frequenz und Rhythmus des Herzschlag (Study of the Frequency and Rhythm of the Hearbeat with a Cardiotachometer)* (Boas and Goldschmidt 1930; Boas and Goldschmidt 1932).[3] The cardiotachometer measured heart rate through electrodes attached to the skin with a harness, and on one page of their book they describe the heart rates of male and female subjects having sex with each other. They found that subjects registered a high heart rate (up to 146 beats per minute) during intercourse, akin to the 150 pulse beats per minute that Martin Mendelsohn found 34 years before (Boas and Goldschmidt 1930). Mary Roach suggests that the subjects were probably Ernst Goldschmidt and his wife (Roach 2008). She states regarding these early experiments, "Rather than risk being fired or ostracized by explaining their unconventional project to other people and trying to press those other people into service, researchers would simply, quietly, do it themselves" (ibid., p. 29). That seems to be the case as well with two German researchers, Hellmuth Klumbies and Gerhard Kleinsorge, physicians specializing in internal medicine in Jena, Germany. Klumbies and Kleinsorge used an electrocardiograph, sphygmograph, and respirator on one man and one woman who fantasized to orgasm (Klumbies and Kleinsorge 1950a; Klumbies and Kleinsorge 1950b). They found that the subjective strength of both the woman's and the man's orgasm correlated in intensity with their increase in blood pressure, respiratory rate, and pulse rate. They hid the descriptions of the machines they used on both subjects by writing them in Latin.[4]

---

[3] A contemporary anonymous reviewer wrote of the English translation of *The Heart Rate:* "The authors have investigated the many factors which influence heart rate, especially the results of emotion, exercise, hot and cold food and the varying conditions of sleep. Particularly instructive are their observations on the morning toilet which in several normal controls raised the pulse rate above 110. The many excellent trackings in the book record the effects of those homely and intimate events of daily life which have escaped scientific study in normal people" (Du B 1933, pp. 579–580).

[4] They concluded that men and women with heart conditions should avoid having orgasmic sex, but realized that cautioning people against having sex was "hopeless," as having unmet sexual needs could be just as much of a strain on the heart as having sex: "Because to forbid coitus is a hopeless start. The physician is probably stronger than Bacchus, but weaker than Venus. Moreover, this is also rejected by well-known authors, because unmet sexuality leads to heightened excitation of the heart—for the heart patients, that should be more harmful than coitus, as one says. Further reason to follow our instructions!" [Denn das coitus verbieten, ist ein

Another example of a multi-machine experiment is that of a Maryland-based researcher in the mid-1950s who aimed to measure the respiratory rate, breath pressure, and breath volume of three married couples during intercourse. Subjects had their noses clamped. The participants wore a mouthpiece with one valve attached to a dry gas test meter and another to a tambour machine that recorded their breath pressure on a kymograph. He found that couples' normal breathing rates tripled and their heart rates were two-and-a-half times higher than a normal resting rate (Bartlett 1956; Roach 2008). A reviewer wrote of this seemingly one-time experiment, "One must admire the heroic performance of R. G. Bartlett's subjects under these conditions of recording" (Zuckerman 1971, p. 309). Roscoe Bartlett, then a twenty-nine-year-old postdoctoral researcher at the National Institutes of Health, presented his research in a packed session of the American Physiological Society in Atlantic City, New Jersey, stating that his intention was "merely to collect facts" (Time 1956).[5] A British experiment that mirrored Bartlett's multi-machine experiments with measuring heterosexual coitus in Britain with a single couple, but found less dramatic changes in breathing rate (Fox and Fox 1969). Other researchers in this era who were interested in mammalian sexual physiology turned to research on animals to explore their interests without stirring up public controversy, but two publicly proclaimed and pursued their interest in human sex research and used machines to do so: Wilhelm Reich and Alfred Kinsey (Gantt 1944).

Wilhelm Reich (1897–1956), best known for inventing the orgone energy accumulator or "orgone box," received his medical degree from the University of Vienna after serving in the Austrian army during World War I. He then trained in and practiced psychoanalysis under Freud (Wilhelm Reich Infant Trust 2011). While in Vienna, he adopted Freud's abandoned ideas that the human sexual libido is actually an energy force in the body, and that the body needed to dispel it properly through orgasm to maintain health (Reich 1927). After he moved to Norway in 1934, he bought an oscillograph (a machine that measured electrical charges), connected it to a kymograph, and began experiments in his office with volunteers. He connected the nipples and genitalia of volunteers to the oscillograph with electrodes, asked them to engage in sex play or to frighten or tickle each other, and then measured the electrical charges from their bodies (Turner 2011). He discovered that orgasm pushed the oscillograph to make the highest waves of any activity, and after he found that a slow-wave "orgone energy" (not

---

(Footnote 4 continued)
hoffnungsloses Beginnin. Der Ärzt ist wohl stärker als Bacchus, aber schwächer als Venus. Au Berdem wird dieses auch von namhaften Autoren abgelehnt, weil unbefriedigte sexualität zu Erregungszustanden des Herzens fürht, die für den Herzkranken—wie man sagt—noch schädlicher sein sollen als der Coitus. Folgen Sie daher unseren Beobachtungen weiter!] (Klumbies and Kleinsorge 1950a, p. 957).

[5] Roscoe G. Bartlett later served ten terms as a member of the United States Congress, representing a district in western Maryland from 1993 until 2013 (Maryland State Archives 2013).

electro-magnetic energy) charged organic materials, he developed the orgone box in the early 1940s to capture that energy for disease prevention and healing (Reich 1937; Bullough 1994). As Reich had moved to the United States in August 1939, his experiments and work fell under the jurisdiction of the U.S. Food and Drug Administration (FDA). Reich and one of his students were convicted of and jailed for contempt of court after they ignored a federal injunction against their research (Wilhelm Reich Infant Trust 2011). The FDA also burned multiple copies of Reich's work, and Reich died in prison of heart failure in 1957. Reich, his orgone box, and a later machine called a cloudbuster, linger in literary, musical, and filmic culture through the present, as many who read about Reich's work believed the orgone box improved a person's sexual and orgasmic potency.[6] Though Reich repeatedly denied that the orgone box heightened sexual prowess, the idea stuck, and Reich is widely remembered as a scientist who attempted to improve people's sex lives via machines.

Alfred Kinsey was an American entomologist-turned-sex researcher who incorporated machines thoroughly into his sex research from its origins. When he began leading a marriage course for undergraduate students at Indiana University, Bloomington, in spring 1938, he found that the students often wanted to meet with him individually to discuss their private sex-related concerns. He soon began to take their sexual histories in a structured interview format, and when the university administration shut down the marriage course in September 1941, Kinsey had already rented punched-card machine equipment to begin processing his collected data (Drucker 2007). As he and his team gathered data from answers to up to 521 questions for each of the eventual 18,000 interviews they made, they transferred the hundreds of thousands of data points onto punched cards in order to combine those points into the tables and charts that made up *Sexual Behavior in the Human Male* (1948) and *Sexual Behavior in the Human Female* (1953) (Kinsey et al. 1948; Kinsey et al. 1953; Drucker 2013). In addition to using punched-card machines, which were at a great distance from the actual acts of human behavior that the holes in the cards represented, he also arranged to have a photographer and filmmaker, Bill Dellenback, make films of sexual encounters privately arranged in Kinsey's home attic (Gathorne-Hardy 1998). Kinsey used the information from those films—some of which he participated in, and some of which he observed—in his discussions of sexual anatomy and physiology in *Sexual Behavior in the Human Female* (Kinsey et al. 1953). The existence of the "attic films," as they were called, were not public knowledge until Kinsey's colleague Wardell B. Pomeroy mentioned them in his biography of Kinsey (Pomeroy 1972). So while Kinsey used machines to process data and to collect it visually, his focus on human sexual behavior and its anatomical, physiological, hormonal, psychological, and neurological aspects eventually earned

---

[6] For example, the Woody Allen film *Sleeper* (1973) prominently features a machine called an Orgasmatron, which Diane Keaton's character Luna Schlosser keeps in her living room. When Luna and Miles Monroe (Erno Windt) want to have sex, they step into the white, tubular machine, it whirrs for a few seconds, and they step out of it, still clothed but looking slightly disheveled.

him a congressional investigation for subversive activities and led to the loss of his research funding (Gathorne-Hardy 1998).

Watson, Reich, and Kinsey were only the most famous examples of scientists who paid a high price for their interest in human sex research that involved machines. If a scientist contemplating working in sex research was to survey the field in the year 1957—a year after Kinsey died, and the year Reich died in prison—that person would see a lot of attention to sexual behavior in certain fields of the human sciences, most prominently psychology, psychiatry, and marriage and family counseling (Davis 2010). That person would see only scattered physiological research, and would certainly notice that the two scientists who used machines to study sexuality were both subject to close scrutiny and political censure, though for different reasons.

This book analyses the ways that machines were incorporated into sex research, and traces how machine-oriented sex research assumed a more central place in academic literature on sexuality. Physicians from the late 1950s through the early 1970s initially used machines in sex research as mechanisms to assert and to enforce medical authority over the bodies of their sexually deviant, and usually male, patients. Then, in a laboratory using volunteers, Masters and Johnson handed a key sex research device—the dildo camera—to female subjects. In the early 1970s and through the present, sex research instruments like the vaginal photoplethysmograph became part of research that aimed to explore, rather than to restrain, sexual physiology and its differences and similarities to subjective arousal. Tracing the history of machines used in sex research shows how the study of physiology became critical to human sex research and to the formation of sexology as a professional academic discipline.

References

Archibald, T. 2005. *Sex machines: Photographs and interviews*. Carrboro, North Carolina: Daniel 13.
Bartlett, R.G. 1956. Physiological responses during coitus. *Journal of Applied Physiology* 9(6): 469–472.
Bercovici, B., and Y. Palti. 1967. Comparison between uterine, upper vaginal, lower vaginal, and digital pulse. *American Journal of Obstetrics and Gynecology* 98(3): 414–418.
Bergner, D. 2009. What do women want? http://www.nytimes.com/2009/01/25/magazine/25desire-t.html?_r=1&. Accessed 13 Apr 2013.
Berry, M.F. 1988. *Why ERA failed: Politics, women's rights, and the amending process of the constitution*. Bloomington: Indiana University Press.
Bijker, W.E., T.P. Hughes, and T.J. Pinch. 1987. Introduction. In *The social construction of technological systems: New directions in the sociology and history of technology*, ed. W.E. Bijker, T.P. Hughes, and T.J. Pinch, 9–15. Cambridge: MIT Press.
Blasius, M., and S. Phelan. 1997. *We are everywhere: A historical sourcebook in gay and lesbian politics*. New York: Routledge.
Boas, E.P., and E.F. Goldschmidt. 1930. *Studien mit dem kardiotachometer über frequenz und rhythmus des herzschlag [Study of the frequency and rhythm of the heartbeat with a cardiotachometer]*. Berlin: Springer.
Boas, E.P., and Goldschmidt. E.F. 1932. *The heart rate*. Charles C. Thomas, Illinois: Springfield.

Brown, M.P. 2000. *Closet space: Geographies of metaphor from the body to the globe.* New York: Routledge.

Bullough, V.L. 1994. *Science in the bedroom: A history of sex research.* New York: Basic Books.

Butler, J. 1990. *Gender trouble: Feminism and the subversion of identity.* New York: Routledge.

Canaday, M. 2009. *The straight state: Sexuality and citizenship in twentieth-century America.* Princeton: Princeton University Press.

Carpenter, D. 2012. *Flagrant conduct: The story of Lawrence v. Texas.* New York: W. W. Norton.

Cohen, D. 2013. *Family secrets: Living with shame from the Victorians to the present day.* London: Viking.

Cook, M. 2007. Queer conflicts: Love, sex and war, 1914–1967. In *A gay history of Britain: Love and sex between men since the middle ages,* ed. M. Cook, 145–178. Oxford: Greenwood World Publishing.

D'Emilio, J., and E.B. Freedman. 1988. *Intimate matters: A history of sexuality in America.* New York: Harper & Row.

Davis, R.L. 2010. *More perfect unions: The American search for marital bliss.* Cambridge: Harvard University Press.

Dickinson, T. 2012. *Mental nursing and "sexual deviation": Exploring the role of nurses and the experience of patients, 1935–1974.* Dissertation: University of Manchester.

Dickinson, R.L., and L.E. Beam. 1931. *A thousand marriages: A medical study of sex adjustment.* Baltimore: Williams & Wilkins.

Dickinson, R.L., and L.E. Beam. 1934. *The single woman: A medical study in sex education.* New York: Waverly Press.

Drucker, D.J. 2007. "A noble experiment": The marriage course at Indiana University, 1938–1940. *Indiana Magazine of History* 103(3): 231–264.

Drucker, D.J. 2013. Keying desire: Alfred Kinsey's use of punched card machines for sex research. *Journal of the History of Sexuality* 22(1): 105–125. doi:10.1353/sex.2013.0002.

Du B, E.F. 1933. Review of The Heart Rate. *American Heart Journal* 8(4): 579–580. doi:10.1016/S0002-8703(33)90705-X.

Faludi, S. 2013. Death of a revolutionary. New Yorker (15 Apr 2013). http://www.newyorker.com/reporting/2013/04/15/130415fa_fact_faludi. Accessed 13 Apr 2013.

Foucault, M. 1977. *Discipline and punish: The birth of the prison.* Trans: A. Sheridan. New York: Vintage.

Foucault, M. 1980. The confession of the flesh. In *Power/knowledge: Selected interviews and other writings, 1972–1977,* ed. M. Gordon, 194–228. New York: Pantheon.

Foucault, M. 1985. *The history of sexuality,* vol. 1. Trans: R. Hurley. New York: Vintage.

Fox, C.A., and B. Fox. 1969. Blood pressure and respiratory patterns during human coitus. *Journal of Reproduction and Fertility* 19(3): 405–415. doi:10.1530/jrf.0.0190405.

Galison, P. 1987. *How experiments end.* Chicago: University of Chicago Press.

Gantt, W.H. 1944. *Experimental basis for neurotic behavior: Origin and development of artificially produced disturbances of behavior in dogs.* New York: Paul B. Hoeber.

Garton, S. 2004. *Histories of sexualities.* New York: Routledge.

Gathorne-Hardy, J. 1998/2004. *Kinsey: Sex the measure of all things; A life of Alfred C. Kinsey.* Bloomington: Indiana University Press.

Gay, P. 1988. *Freud: A life for our time.* New York: W. W. Norton.

Geer, J.H., P. Morokoff, and P. Greenwood. 1974. Sexual arousal in women: The development of a measurement device for vaginal blood volume. *Archives of Sexual Behavior* 3(6): 559–563. doi:10.1007/BF01541137.

Gerhard, J.F. 2001. *Desiring revolution: Second-wave feminism and the rewriting of American sexual thought, 1920 to 1982.* New York: Columbia University Press.

Grosz, E. 1995. *Space, time, and perversion: Essays on the politics of bodies.* New York: Routledge.

Henson, D.E., et al. 1977. Temperature change of the labia minora as an objective measure of female eroticism. *Journal of Behavior Therapy and Experimental Psychiatry* 8(4): 401–410. doi:10.1016/0005-7916(77)90011-8.

Kinsey, A.C., W.B. Pomeroy, and C.E. Martin. 1948. *Sexual behavior in the human male*. Philadelphia: W. B. Saunders.

Kinsey, A.C., et al. 1953. *Sexual behavior in the human female*. Philadelphia: W. B. Saunders.

Klumbies, G., and H. Kleinsorge. 1950a. Das herz in orgasmus [The heart in orgasm]. *Medizinische Klinik* 45(31): 952–958.

Klumbies, G., and H. Kleinsorge. 1950b. Circulatory dangers and prophylaxis during orgasm. *International Journal of Sexology* 4: 61–66.

Latour, B., and S. Woolgar. 1986. *Laboratory life: the construction of scientific facts*. Princeton: Princeton University Press.

Lefebvre, H. 1991. *The production of space*. Trans: D. Nicholson-Smith. Oxford: Blackwell.

Levins, H. 1996. *American sex machines: The hidden history of sex at the U.S. patent office*. Holbrook, Massachusetts: Adams Media.

Lieberman, H. 2011. Why miss the super-pleasures of life? Consumer shaping of the electromechanical vibrator. Unpublished conference presentation, Society for the History of Technology, Cleveland, Ohio.

Magoun, H.W. 1981. John B. Watson and the study of sexual behavior. *Journal of Sex Research* 17(4): 368–378.

Maines, R.P. 1999. *The technology of orgasm: "Hysteria," the vibrator, and women's sexual satisfaction*. Baltimore: Johns Hopkins University Press.

Mancini, E. 2010. *Magnus Hirschfeld and the quest for sexual freedom: A history of the first international sexual freedom movement*. Basingstoke: Palgrave Macmillan.

Mansbridge, J.L. 1986. *Why we lost the ERA*. Chicago: University of Chicago Press.

Marcus, E. 2002. *Making gay history: The half-century fight for lesbian and gay equal rights*. New York: Harper.

Maryland State Archives. 2013. Roscoe G. Bartlett. http://msa.maryland.gov/msa/mdmanual/39fed/06ushse/former/html/msa01956.html. Accessed 21 Mar 2013.

Masters, W.H., and V.E. Johnson. 1966. *Human sexual response*. Boston: Little, Brown.

Max Planck Institute for the History of Science, Berlin. 2008–2010. The virtual laboratory: Technology. http://vlp.mpiwg-berlin.mpg.de/technology. Accessed 10 Mar 2013.

McLaren, A. 1999. *Twentieth-century sexuality: A history*. Oxford: Blackwell.

McLaren, A. 2007. *Impotence: A cultural history*. Chicago: University of Chicago Press.

Mendelsohn, M. 1896a. Ist das radfahren als eine gesundheitsgemässe übung anzusehen und aus ärztlichen gesichtspunkten zu empfehlen? [Is bicycling considered a healthy exercise and recommended as a focus for medical health?]. *Deut Med Wochenschr* 22: 277–278.

Mendelsohn, M. 1896b. Ist das radfahren als eine gesundheitsgemässe übung anzusehen und aus ärztlichen gesichtspunkten zu empfehlen? [Is bicycling considered a healthy exercise and recommended as a focus for medical health?]. *Deut Med Wochenschr* 22: 300–301.

Mendelsohn, M. 1896c. Ist das radfahren als eine gesundheitsgemässe übung anzusehen und aus ärztlichen gesichtspunkten zu empfehlen? [Is bicycling considered a healthy exercise and recommended as a focus for medical health?]. *Deut Med Wochenschr* 22: 333–334.

Mendelsohn, M. 1896d. Ist das radfahren als eine gesundheitsgemässe übung anzusehen und aus ärztlichen gesichtspunkten zu empfehlen? [Is bicycling considered as a healthy exercise and recommended as a focus for medical health?]. *Deut Med Wochenschr* 22: 366–368.

Mendelsohn, M. 1896e. Ist das radfahren als eine gesundheitsgemässe übung anzusehen und aus ärztlichen gesichtspunkten zu empfehlen? [Is bicycling considered as a healthy exercise and recommended as a focus for medical health?]. *Deut Med Wochenschr* 22: 381–384.

Mendelsohn, M. 1896f. Ist das radfahren als eine gesundheitsgemässe übung anzusehen und aus ärztlichen gesichtspunkten zu empfehlen? [Is bicycling considered a healthy exercise and recommended as a focus for medical health?]. *Deut Med Wochenschr* 22: 398–401.

Mendelsohn, M. 1901. *Der ausbau im diagnostischen apparat der klinischen medizin [Developments in the diagnostic apparatus of clinical medicine]*. Wiesbaden: Bergmann.

Milk, H. 2012. The Harvey Milk interviews: In *his own words*, ed. V. Emery San Francisco: Vince Emery Productions.

Oosterhuis, H. 2000. *Stepchildren of nature: Krafft-Ebing, psychiatry, and the making of sexual identity*. Chicago: University of Chicago Press.

Oosterhuis, H. 2012. Sexual modernity in the works of Richard von Krafft-Ebing and Albert Moll. *Medical History* 56(2): 156–183. doi:10.1017/mdh.2011.30.

Palti, Y., and B. Bercovici. 1967. Photoplethysmographic study of the vaginal blood pulse. *American Journal of Obstetrics and Gynecology* 97(2): 143–153.

Pinch, T.J., and W.E. Bijker. 1987. The social construction of facts and artifacts: Or how the sociology of science and the sociology of technology might benefit each other. In *The social construction of technological systems: New directions in the sociology and history of technology*, ed. W.E. Bijker, T.P. Hughes, and T.J. Pinch, 17–50. Cambridge: MIT Press.

Pomeroy, W.P. 1972. *Dr. Kinsey and the Institute for Sex Research*. New York: Harper & Row.

Power, L. 1995. *No bath but plenty of bubbles: An oral history of the Gay Liberation Front, 1970–1973*. London: Cassell.

Reich, W. 1927. *Der funkton des orgasmus [The function of the orgasm]*. Leipzig: Internationaler Psychoanalyticer Verlag.

Reich, W. 1937. *Experimentelle ergebnisse über die elektrische funktion von sexualität und angst [Experimental results about the electrical function of sexuality and anxiety]*. Copenhagen: Sexpol Verlag.

Roach, M. 2008. *Bonk: The curious coupling of sex and science*. New York: W. W. Norton.

Sauerteig, L.D. 2012. Loss of innocence: Albert Moll, Sigmund Freud and the invention of childhood sexuality around 1900. *Medical History* 56(2): 156–183. doi:10.1017/mdh.2011.31.

Scott, J.C. 1930. Systolic blood-pressure fluctuations with sex, anger, and fear. *Journal of Comparative Psychology* 10: 97–114. doi:10.1037/h0073671.

Sedgwick, E.K. 1990. *Epistemology of the closet*. Berkeley: University of California Press.

Shilts, R. 1988. *The mayor of Castro Street: The life and times of Harvey Milk*. New York: St. Martin's Press.

Sigusch, V. 2012. The sexologist Albert Moll—between Sigmund Freud and Magnus Hirschfeld. *Medical History* 56(2): 184–200. doi:10.1017/mdh.2011.32.

Soja, E.W. 1989. *Postmodern geographies: The reassertion of space in critical social theory*. London: Verso.

Staudenmaier, J.M. 1994. Rationality versus contingency in the history of technology. In *Does technology drive history?*, ed. M.R. Smith, and L. Marx, 259–274. Cambridge: MIT Press.

This American Life. 2002. 81 words. http://www.thisamericanlife.org/radio-archives/episode/204/81-words. Accessed 12 Mar 2013.

Time Magazine. 1956. Medicine: Wired for love. http://www.time.com/time/subscriber/article/0,33009,824206,00.html. Accessed 21 Mar 2013.

Turner, C. 2011. *Adventures in the orgasmatron: How the sexual revolution came to America*. New York: Farrar, Straus, and Giroux.

Vance, C.S. 1984. *Pleasure and danger: Exploring female sexuality*. London: Routledge & Keegan Paul.

Watson, J.C. 1913. Image and affection in behavior. *Journal of Philosophy, Psychology and Scientific Methods* 10(16): 421–428.

Wilhelm Reich Infant Trust. 2011. Biography of Wilhelm Reich. http://www.wilhelmreichtrust.org/biography.html. Accessed 10 Mar 2013.

Zuckerman, M. 1971. Physiological measures of sexual arousal in the human. *Psychological Bulletin* 75(5): 297–329. doi:10.1037/h0030923.

Chapter 2
The Penile Strain Gauge and Aversion Therapy: Measuring and Fixing the Sexual Body

Abstract The early Cold War had far-reaching effects into the gendered and sexual aspects of adult life. Those effects are clearly evident in the United States and United Kingdom, in which physicians and their assistants used technology to practice aversion therapy on male patients who were sexually "deviant," including homosexuals, cross-dressers, fetishists, sadists, and masochists. They designed a form of aversion therapy using an electroshock machine together with a penile strain gauge in order to shock those men into heterosexually normative sexual desires and behaviors. The treatment did not work, and physicians abandoned it in the light of the rise of gay rights and amid civil rights movements and professional reinterpretations of homosexuality in the late 1960s and early 1970s. This chapter first describes the setup of early voluntary sex research laboratories, which provided a framework for the setup of laboratories using involuntary subjects. It then details the theoretical framework, based on Martha Nussbaum's conception of objectification and Michel Foucault's ideas on the functions of institutions in policing people's bodies and minds. It describes how aversion therapy (originally using drugs) and the use of electricity on human and animal bodies came together in the mid-1960s in the form of electroshock aversion therapy. That description includes primary-source interviews with patients, doctors, and nurses, all of whom had varying responses to the use of such therapy. The chapter concludes with an explanation of how electroshock aversion therapy largely ended by the mid-1970s and how some doctors, nurses, and patients resisted it.

Keywords Electroshock · Aversion therapy · Penile strain gauge · Homosexuality · Sex research · Cold War

The Cold War in the United States and United Kingdom, as multiple historians have discussed, was a time of gender and sexual policing by individuals, social groups, and mass media culture. It was also a time of medical, and particularly

An erratum to this chapter is available at 10.1007/978-94-007-7064-5_6 psychiatric, authority (Johnson 2004; Cuordileone 2005; Cook 2007; Cohen 2013). Numerous individuals, particularly homosexual men, but also lesbian women, transsexuals, transgendered individuals, cross-dressers, sadomasochists, fetishists, and pedophiles were at the margins of society. Medical, legal, and political authorities attempted to change them to conform to societal norms, to push them into changing themselves, and to punish them for nonconforming. The laboratory settings and machines used in sex research during the 1950s and 1960s were often used in the service of such agendas.

As machine-based physiological research in the 1950s and 1960s focused largely on studying or fixing men, researchers oriented and developed their machines mostly for male bodies. The gender and sexual values of the Cold War both supported and were reflected in the sex research done at the time. However, scientists also established broader frameworks and practices for the use of machines in sex research that set the stage for studies that would ultimately be more liberating than restrictive. Human sex research using machines developed in two main directions in the 1950s and 1960s: one in which individual, largely isolated researchers invented and experimented with various types of machines to pursue their interests in sexual physiology. The second avenue of research used machines as tools to normalize men considered "deviant" by cultural, medical, and legal standards. (The third avenue—comparative research with individuals and couples—is the subject of Chap. 3.) Those two early paths of sexual physiology research illustrated two approaches to the role of machines in sexual medicine: one aimed at the discovery of basic physiology, the other toward regulation and punishment. This chapter has four parts—first, a brief overview of the gender and sexual issues in the 1950s and 1960s in the U.S. and U.K.; second, a description of the laboratories and machines used in exploratory research; third, the machines and practices used in aversion therapy; and fourth, resistance tactics that medical staff and patients used to subvert aversion therapy. This chapter illustrates that machines used in sex research were not neutral: they became loci for scientific or self-discovery, fear, pain, resistance, and many other sex-related experiences and emotions.

Historical Background

Elaine Tyler May, in her now-classic book *Homeward Bound,* describes how ideals of containment in the United States intersected nuclear and foreign policies of the Cold War with popular constructions of the white, middle-class, suburban, heterosexual, single-family home (May 1988). She argues that American governmental policies directed toward containing the threats of nuclear energy, the atom bomb, and Communism also included medical and legal policies corralling dangerous social forces so that people could live secure, fulfilling lives. Such social forces prominently included any kind of sexual deviance, so people (often unhappily, as May argues) contained, or tried to contain, themselves as well. They attempted to police their own bodies as Michel Foucault described: to conform inwardly and outwardly to the social mores of the time (Foucault 1977). Legal and medical establishments in the United States and United Kingdom then took on the task of containing people who were unwilling or unable to contain themselves, and those people were willingly or unwillingly subjected to the power of machines to eliminate the deviant parts of themselves. In addition, an expert culture surrounding marriage was rising, with more and more individuals claiming science-based advice and formulas for marital happiness and success (Minton 1988; Davis 2010; Lewis 2010). Such expertise was often based on the premise that heterosexual marriage was the foundation of a strong American nation, and any departure from that ideal was a threat to national stability (Reumann 2005). May's, Davis's, and Reumann's work also shows the importance of postwar expert culture in enforcing the rules of heterosexual containment, a culture which was similar to but not identical to the sociocultural situation in Britain.

American cultural ideals emphasized domestic stability, security, and the nuclear family following the Depression and the end of World War II. The language of American Cold War politics emphasized the importance of heterosexual masculinity, as the strength of the nation was rhetorically embedded in the physical and moral strength of its adult men. Cold War political culture "put a new premium on hard masculine toughness and rendered anything less than that soft and feminine and, as such, a real or potential threat to the security of the nation" (Cuordileone 2000, p. 516). Homosexual or queer men embodied that particular threat. Historians including John D'Emilio, Estelle Freedman, and David Johnson have written of the heightened tension in postwar sexual politics, particularly the fears of a "homosexual menace" and an accompanying anxiety that homosexuals were aligned with, or tools of, the Communist party (D'Emilio and Freedman 1988; Johnson 2004; Drucker 2010). As Freedman and D'Emilio put it, "the labeling of homosexuals as moral perverts and national security risks, along with the repressive policies of the federal government, encouraged local police forces across the country to harass them with impunity" (D'Emilio and Freedman 1988, p. 293). In a political and cultural atmosphere largely hostile to homosexuality, some physicians took the opportunity to use their medical authority to treat the non-heteronormative nature of men (and some women) as a disease that they could cure in a laboratory setting.

Following World War II in Britain, there was a public cultural and political emphasis on the stability of family traditions and gender roles. Nonetheless, plenty of signs existed that the values and the traditions of the Victorian, heterosexual, Christian, and reproductive family were fragmenting in the minds of the broader population beginning in the 1950s. By the beginning of the 1960s, Callum G. Brown suggests that British Christian women and men began to feel alienated from church teachings, were less likely to attend religious services, and more likely to engage in premarital intercourse than in the previous decade (Brown 2011). Frank Mort emphasizes the rise in public expressions of scientific expertise in sexual matters and the wider visibility of gay male sexuality and non-marital sexuality widely—along with figures such as "the man about town" and concepts such as the rise of the "pleasure economy" in the 1960s (Mort 2010). After the *Wolfenden Report* appeared in 1957, many physicians saw an opportunity to experiment with using medicine to help people conform to contemporary societal norms (Wolfenden et al. 1957).

The late 1950s through the late 1960s were also opportune years for the British medical establishment to study and perhaps to "fix" homosexuality. After the conviction for "gross indecency" and death by apparent suicide of the scientist Alan Turing, among other widely publicized criminal trials for homosexuality-related offenses, the Conservative government set up a Departmental Committee in 1954 under the leadership of Sir John Wolfenden, vice chancellor of Reading University, to consider homosexual offenses along with prostitution (Daily Mail 2007; Bryant 2012). The *Report of the Departmental Committee on Homosexual Offences and Prostitution* (popularly known as the *Wolfenden Report)*, after multiple hearings, interviews, and debates, was finally published in 1957. It recommended the decriminalization of homosexual sex in private between consenting adults over the age of 21; that buggery (sodomy) be reclassified from a felony to a misdemeanor criminal offense; that estrogen treatment should be available to any man who wanted it; and that physicians should continue to research the causes of—and treatments for—homosexuality (Dickinson 2012, p. 110). The committee concluded that homosexuality was not a disease, but that men who wanted medical treatment should be able to seek it. The report did not consider lesbian women, as sex between women was not a crime in the U.K. Treatment aims could involve a change of direction in sexual preference, a better adaptation to life in general, and greater continence and self-control (Wolfenden et al. 1957, p. 66). According to the committee, homosexuals suffered from a "maladjustment to society," and their treatment could involve estrogen, psychotherapy, help from probation officers, and counseling from clergy (ibid., pp. 66–68).

In the decade between the *Wolfenden Report* and the Sexual Offences Act of 1967 (when male homosexuality was partially decriminalized), many British physicians, particularly psychologists and psychiatrists who felt as though the *Wolfenden Report* gave them a mandate to do so, took it upon themselves to experiment with cures on homosexual men. As Tommy Dickinson put it, "following Wolfenden, there was a distinct altering of notions regarding homosexuality from a criminal perspective to understandings of the subject as pathology. There was a shifting of control and power from the courts to the medical profession, many of whom were optimistically promoting their worth in being able to cure these individuals by reporting successful outcomes" (Dickinson 2012, pp. 133–34). The *Wolfenden Report* helped establish the cultural idea that scientific experts could cure social maladies, as embodied in British citizenry, and that the government had faith that they could do so. One physician hinted that medicine's inability to find a simple solution to homosexuality led to frustration and anger among medical practitioners, and that those feelings may have been an inspiration to test more extreme solutions. Basil James introduced a 1962 article by describing the failure of psychoanalysis to cure homosexuality as "the feeling of therapeutic impotence which the practitioner so often feels when faced with the problem of homosexuality" (James 1962, p. 768). By the late 1950s, in both the U.K. and U.S., the social and political context for sexual physiology experiments was in place. First, however, physicians needed the right spaces and tools for the job: properly organized research laboratories with machines to measure the physiological processes that interested them the most. Chief among those was a penile measurement instrument.

The Sex Research Laboratory

The physical space of the laboratory, the placement of instruments and machines within it, and the ways that researchers' and subjects' bodies moved within the laboratory all affected how the research took place, how researchers understood their own investigative practices, and the researchers' results (Latour and Woolgar 1986). Understanding the meaning of laboratory space and its effects on research outcomes is particularly important for analyzing studies of human behavior, as researchers' and subjects' interactivity within the laboratory environment frames how individuals behaved throughout the research process. The laboratory space, its instruments, and its machines also contained or placed limitations on the possible outcomes of the research. The positions of the researchers (inside the laboratory or viewing from outside), the positions and clothing of the subjects, the furniture, audio or video recording equipment, lighting, and temperature of the room all affected the ability of subjects to perform, and the data that the researchers hypothesized that they would find. Close analysis of the space of the human sex research laboratory, along with the actors, the instruments, and practices, shows that laboratory settings provided subjects with some privacy and anonymity but also reinforced certain kinds of sexual behaviors and physiological responses as the only aspects of "normal" sexuality.

Sex research laboratories across the Western world were often remarkably alike whether they were used to study or to change people. Sometimes scientists were simply testing machines to see what tools might work most effectively in the future. Others thought more broadly, considering what just the right set of machines and experiments could identify about the bodily mechanisms of arousal and desire, and what combination of forces in body and mind led to arousal, response, and orgasm. The spaces and instruments of the laboratories, along with the interactions of the researchers with the spaces and instruments, placed constraints on the possibilities for research, regardless of the perspectives and intentions of the researchers themselves. While some furniture and machines permitted a relatively wide range of movement and thought, others restricted the actions of the participants so that a realistic measure of their "natural" behavior was limited.

Past laboratory spaces become visible in the present through the textual descriptions in articles from psychological and other medical journals; from some second-hand descriptions by biographers of sex researchers; and researchers' and participants' firsthand accounts of their experiences. Few of the researchers from the earliest period of machine-based sex research described their laboratory spaces in much detail. Research often took place at hospitals or science laboratories of universities, and most study subjects were college students, patients under the researcher's care, local volunteers solicited through billboard postings, or those whom Alfred C. Kinsey euphemistically called for his work "friends of the research": acquaintances, family members, and friends of the lead investigators who were interested in sex, were curious about the research process, and wanted to help the project (Gathorne-Hardy 1998/2004, p. 218). The voluntary research subject, who was usually only given a vague outline of what he or she was about to see and to do, was shown to a small room outfitted with a chair, a slide or film projector, a projection screen, wall outlets, and the measurement devices. Sometimes researchers would be seated behind a one-way mirror and be able to watch the subject; other times they would be in the room watching the patient from behind or audio- or tape-recording the entire proceedings.

Researchers used a wide variety of devices to see if any of them could become standardized equipment. The use of machines moved through phases: as one researcher would seem to get promising results from a new invention, others would try it and get mixed or poor results, and then the machine would disappear. The nascent research community would affirm the success of the few that seemed to deliver results by incorporating them repeatedly into experiments. Whatever devices researchers decided to use, sex research demanded a level of physical intimacy between investigator and subject. The researchers needed to fit all of the devices properly and to instruct the subject in how to use and to adjust the machines if necessary. The most widely used machines for both male and female subjects were respirators to measure breathing, small blood pressure cuffs that fit on fingers, toes, or wrists, electrocardiograms to measure heart rate, electroencephalograms to measure brain activity, and galvanic skin transducers, which measured the levels of sweat or dryness on the skin. The subject was hooked up to the electrocardiograms and electroencephalograms with electrodes. Particular machines were also designed for genitals. For men, the most common device was a mechanical or mercury-in-rubber penile strain gauge, while one researcher developed a scrotal strain gauge that she used in a similar manner (Zuckerman 1971, p. 314; Bell 1975; Bell and Stroebel 1973). The most common machine for women was the vaginal photoplethysmograph (Chap. 4).

The penile strain gauge was initially developed in Tübingen, Germany, in 1936 to measure men's erections in sleep. Researchers at the University of Tübingen measured the number of erections of a single sleeping man overnight using a metal cuff attached to his penis. Circumferential penile fluctuations were recorded on a kymograph through a magnetic recorder (Ohlmeyer et al. 1944). The metal cuff only measured the presence or absence of an erection, not any specific measurements of penile circumference or rigidity. Ohlmeyer and colleagues later repeated the initial experiment with six more men (Ohlmeyer and Brilmayer 1947). This device inspired other inventions for measuring erection, including a mercury-in-rubber strain gauge.

The late 1940s and early 1950s was a time of innovation in the American development of strain gauge technology generally. As the authors of *The Strain Gauge Primer* put it in 1955, "only the last decade has seen outstanding advancement in the art of strain management" (Perry and Lissner 1955, p. 1). In 1949, the American researcher R. J. Whitney adapted the strain gauge to measure muscular tension in a subject's arm or leg, using mercury enclosed in a rubber tube attached to a Wheatstone bridge circuit. As he wrote in his initial description of it, "the band... forms a mercury-in-rubber resistance strain gauge, and extension or shortening of the band will give a corresponding increase or decrease in its electrical resistance" (Whitney 1949, p. 6). This measure had great sensitivity, and Whitney intended other doctors to use it to measure blood flow and limb volume in relation to water intake and to record muscular movements. However, the Whitney strain gauge was not adapted to medical use for sexual investigations right away. Beginning in the mid-1960s, though, researchers modified the Whitney strain gauge to measure the change in penile circumference in millimeters (Gibbons et al. 1963; Barlow et al. 1970; Britt et al. 1971; Farkas et al. 1979). In the late 1960s and early 1970s, other researchers modified the original German metal penile strain gauge so that it too could measure penile circumference (Laws and Rubin 1969; Laws and Pawlowski 1973; Laws and Bow 1976; Laws 1977). Both the mercury-in-rubber strain gauge and the mechanical strain gauge remained in use through the mid-1980s, when an electromechanical strain gauge was added to the market of penile measurement devices.

Researchers from the late 1950s onward also used a penile vacuum pump, which measured the displacement of air in a sealed tube placed over the penis. Penile vacuum pumps to restrict blood flow were available as early as the 1860s in France, and they were a near-constant part of quack medical practitioners' toolkits as impotence cures (Seraine 1865; McLaren 2007). Kurt Freund in Czechoslovakia developed a mechanized vacuum device in order to test men who claimed to be homosexual as a means of getting out of the country's two-year compulsory military service requirement (Freund 1957; Freund et al. 1958; Freund 1963; Freund 1965; Chivers and Bailey 2007).[1] Freund believed that a change in penile volume over a certain number of milliliters indicated that the subject was aroused. Physicians could hook the vacuum pump to men who were shown images of men and women in different stages of undress to test for homosexuality, a desire to cross-dress, or pedophilia (McConaghy 1967; Rosen 1973; Rosen et al. 1975; Rosen and Kopel 1977; Rosen and Keefe 1978). Freund wrote toward the end of his career that an unnamed colleague's work on finger plethysmography had inspired his work on penile plethysmography, but he did not detail the process of creating the vacuum pump (Freund 1991). Of course, Freund and the other researchers described below were proceeding on the assumption that an erection meant only one thing: sexual

---

[1] Even though homosexuality was illegal in Czechoslovakia, and admitting to homosexuality could land them in treatment programs designed to make them heterosexual, nevertheless some men preferred to lie about their sexual preferences and go through a treatment program rather than serving in the military (Freund 1991).

response, even though Kinsey's and his younger colleague Glenn V. Ramsey's research in the 1940s had shown that men's and boy's erections could be responses to sexual and non-sexual stimuli alike, including friction with clothing, taking a shower, watching a war movie, riding in an airplane, or anxiety (Kinsey et al. 1948; Ramsey 1943). Even though erections could indicate a wide range of emotions, researchers considered them only sexually (Fig. 2.1).[2]

Researchers often showed subjects a set of still or moving images to measure their response: usually a "control" image of landscapes or other nature scenes, some type of sex-related scene that involved varying degrees of nudity (often stills from *Playboy* magazine for men), and a third set of images designed to invoke horror, fear, or disgust. Some researchers also used audiotapes of erotic and non-erotic stories or gave subjects material to read and to then fantasize about, but listening and reading were much less popular. The images or films were often shown in a random order, and subjects were usually given some time to rest in between image sets. The aim behind using horrific images was to compare bodily processes like heart, breath, and pulse rates with different kinds of stressors. Some used images of skin diseases, such as ulcerated legs, but the most common horror images, either film or still, were of Holocaust concentration camps showing living or deceased prisoners (Zuckerman 1971; Wood and Obrist 1968; Bernick et al. 1971). There exists no record of subject reactions to such an extreme juxtaposition of images in a laboratory setting, sometimes without air conditioning or room temperature control, with strangers observing them, being partially or fully nude, and hooked up to at least two machines. Sometimes they were given a small stipend, but probably not enough to offset the strangeness of their experiences. Researchers were often testing what machines would best measure arousal, and while they generally expected heterosexual responses from their oddly posed subjects, sometimes subjects defied those assumptions.

For example, two researchers at Chatham College in Pittsburgh, Pennsylvania, in 1965, used a galvanic skin transducer to study sexual stimuli in twenty women who were seniors in college. The women were attached to the transducer, which fit around the hand and finger, and then were shown images of men and women in various stages of undress. They found that women had the highest skin temperature change when viewing nude still images, whether the images were of men or women. They concluded, "whether or not the responses measured in this study reflected guilt, shock, disgust, or sexual eroticism...the responses may be said to be sexual at least in part" (Loiselle and Mollenauer 1965, p. 277). So subjects sometimes validated and occasionally challenged the heteronormative assumptions of researchers. Most (those not involved in aversion therapy, that is) were generally willing to accept and to include same-sex arousal in their results—if they

---

[2] Interestingly enough, no other scientist has ever replicated Ramsey's research on the different possible meanings of erections. Thus scientists generally presume that the erections they measure indicate sexual response, when the erections also could indicate a reaction of fright, anxiety, or other emotions. John Bancroft pointed out that lack of research at a conference in 2003 (Bancroft in Janssen (ed) 2007, p. 308).

Fig. 2.1 Volumetric penile transducer. Freund et al. (1965), p. 170, Fig. 2. Copyright 1965 by the Society for the Experimental Analysis of Behavior, Inc

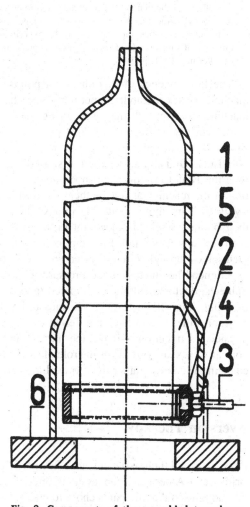

Fig. 2. Components of the assembled transducer.
1 glass cylinder
2 plastic ring
3 metal tube with threads
4 locknut
5 rubber cuff
6 flat soft sponge rubber ring

saw it. It is even clearer that researchers did not know what kinds of techniques and equipment could accurately isolate and then measure something as elusive as sexual arousal and response, let alone what instrument could find the physiological reasons for those phenomena.

A reviewer of machine-based sex research summarized the experience of its subjects this way:

> The effects of the general experimental situation on the subjects have not been considered in most experiments….Even physiological responses may be influenced…by instructions, or the characteristics and behavior of the experimenter….Failure to consider the human qualities of subjects can often lead to erroneous conclusions in psychological experiments (Zuckerman 1971, p. 326).

The last sentence of this quotation points to a crucial element in the use of machines in sex research. Sometimes researchers were so interested in testing out a machine, the fact that they were using machines on actual human subjects who could feel discomfort or even just puzzlement seemed secondary. Such was the case with the original electroconvulsive therapy (ECT) machine, for example, as after Lucio Bini and Ugo Cerletti first used it in Rome in April 1938, "experiments with the Bini–Cerletti electroshock apparatus continued, as much for the perfection of the machine as for the therapy" (Aruta 2011, p. 412). As researchers were trying to learn more about physiology, their subjects were newly learning how their bodies as sexual objects interacted with machines. Perhaps participating in such research was a lark for them, or perhaps they learned something new about themselves through an unexpected erectile response to a nude same-sex image or experienced heightened blood pressure when hearing an erotic story read aloud. Subjects, doctors, and nurses involved in aversion therapy also learned new ways of being, interacting, testing, and measuring the sexual body. Their interactions with machines would have a different cast than purely voluntary research, as the physician's intentions—if not the nurses' and the patients'—were to fix a non-normative sexual body. Thus the role of machines in this type of research shifted to changing the body, rather than to exploring it.

Aversion Therapy

Physicians and their assistants were key participants in the medical treatments that British and American court systems offered sexual deviants instead of jail time. "Although some participants chose to undergo treatments instead of imprisonment or were encouraged through some form of medical coercion, most were responding to complex personal and social pressures that discouraged any expression of their sexuality" (Smith et al. 2004, p. 427). Doctors involved in aversion therapy from the mid-1950s through the mid-1970s thought that machines were especially effective, because they permitted distance between doctor and patient, unavoidable in nausea drug–based aversion therapy. Electromechanical shock machines too, seemed quicker and more effective, but in fact aversion therapies only rarely worked.

Cold War masculinity was linked specifically to technology, as shown by the intersected use of the strain gauge with the electroshock machine to police male sexuality. While the strain gauge measured erection, the electroshock machine, in ways that hormonal or medicinal treatments for aversion therapy could not, gave medical practitioners the perception that they could retrain a man's sexuality through the use of seemingly invisible techniques. Put together, these devices would contain, constrain, and shape Cold War era masculinity into a socially acceptable form. Containment of bodily desire would lead to containment of personal and national risk (Cuordileone 2005). Anyone who did not conform to such masculine ideals could be, and often was—to borrow the title of Foucault's well-known work—disciplined and punished (Foucault 1977). Machines permitted researchers to treat patients like objects that needed fixing, not subjects with autonomy. However, there was room for male subjects (and sometimes for their nurses) to subvert these disciplines and to use them against the medical communities' attempts to change them.

Aversion therapies were well-known treatments in the Cold War era, especially as a cure for alcoholism. By the early 1960s, however, they had become popular with physicians seeking to cure homosexual desire and behavior and to replace it with heterosexual desire and behavior. The now-classic 1971 film, *A Clockwork Orange,* depicts a nightmarish version of aversion therapy called the Ludovico Treatment used on the main character, Alex, who is in prison for murder. He is placed in a straightjacket, his eyes are held open with a machine while a technician drops liquid into his eyes, and doctors show him films with Nazis marching and destroying buildings while his favorite Beethoven music plays over the top. At one point he screams, "Stop it, stop it, please I beg you." The male doctor replies, "I'm sorry, Alex. This is for your own good. You have to bear with us for a while." Alex screams that he has seen the light and is cured, but the Nazi films continue in front of his artificially opened eyes. The idea that a patient who wanted to be cured of a particular desire must associate his most beloved feelings and experiences with atrocities was the logic of aversion therapy.

The two theoretical works that frame the following discussion of aversion therapy are Foucault's *Discipline and Punish* and Martha Nussbaum's "Objectification" (Foucault 1977; Nussbaum 1995). Foucault describes how officials in the military, prison, educational, and hospital institutions of late eighteenth-century France began to exert control over the movement of bodies. "The human body was entering a machinery of power that explores it, breaks it down, and rearranges it." That "mechanics of power" determined "how one may have a hold over others' bodies, not only so that they may do what one wishes, but so that they may operate as one wishes" (Foucault 1977, p. 138). Elizabeth Grosz furthers Foucault's emphasis on the importance of hospitals' powers over unruly bodies, pointing out that "the increasing medicalization of the body...demonstrate[s] a body pliable to power, a *machinic* structure in which 'components' can be altered, adjusted, removed, or replaced. The body becomes increasingly regarded as functional" (Grosz 1995, p. 35, emphasis in original). Institutional control over the physical movement of bodies, particularly using machines, and the internal processes that governed their actions thus literally reshaped the body politic. Foucault's description of how disciplinary powers create "docile bodies" elides neatly with Nussbaum's seven-part breakdown of the ways that humans can objectify others. Nussbaum's theory shows how aversion therapy could objectify a patient using a seven-part concept of objectification: instrumentality, denial of autonomy, inertness, fungibility, violability, ownership, and denial of subjectivity (Nussbaum

1995, p. 257). Using that breakdown of objectification, particularly focusing on the denial of autonomy and subjectivity, shows that from the 1950s through the 1970s, even innocuous-seeming machines such as film projectors and tape recorders served as tools for medical practitioners to deny homosexual and other so-called deviant persons their autonomy and subjectivity.

Well before electroshock became part of aversion therapy, there was a long history of people using machines to police the human body. With the development of widely available commercial electricity networks in urban areas in the 1880s came the use of electricity on human and animal bodies. The most obvious and visible type of electricity applied directly to human bodies was the electric chair, seen in highly publicized photographs in the 1920s, and particularly in the photograph of the American convicted murderer Ruth Snyder's death by electrocution in 1928 (Shahid 2008). The first mention of electroshock therapy for "curing" homosexuality in English was a short notice in the *Psychological Journal* in 1935. That article had very little detail, as it was only a paragraph-length record of a conference paper, but in it the psychologist Louis Max stated that he gave a man electric shocks in conjunction with his homosexual fantasies, aiming to cause "a diminution of the emotional value of the sexual stimulus" (Max 1935, p. 734). After three months of treatment, the man claimed to be "95 per cent cured" (ibid.). Electroconvulsive therapy, as briefly mentioned above, was first developed in Italy in the late 1930s to treat severe depression and schizophrenia (Dickinson 2012, pp. 148–149; Aruta 2011). The patient was given a strong shock with electricity directly applied to the head, which induced an immediate seizure and often unconsciousness. Several members of a hospital's medical staff needed to hold the patient down to ensure that he or she did not self-injure or fall off the bed or examination table. Aversion therapy for homosexuality, by contrast, used a much lower amount of electricity applied to the arm or leg that caused a shock through the limbs but did not cause the seizure and unconsciousness that ECT did. As physicians continued to use ECT to treat severe mental illness, others began to apply smaller amounts of electricity in aversion therapies in order to cure mental conditions they considered problematic but less debilitating, including homosexuality.

It was in the mid-1950s, at a peak of the Cold War internationally, that books and articles were published describing experiments with aversion therapies, usually for alcoholism, but also for smoking, fetishes (such as for shoes, lingerie, or baby carriages), "obsessional ruminations," and even for writer's cramp (Raymond and O'Keeffe 1965; Liversedge and Sylvester 1955; Hilgard 1961; Raymond 1956; Russell 1970). The only machines used at this point were audiotapes, and sometimes films or projected still images. Most aversion therapies were conducted using a classical conditioning technique, in which therapists attempted to get the patient to associate his or her behavior with nausea, physical pain, and the idea that the behavior was truly sick and needed to be abandoned (Rachman 1961). Physicians or their assistants gave patients a nausea drug at two- to three-hour intervals, deprived them of sleep, and often woke them to view still images or to listen to an audiotape about the terribleness of their behavior and the goodness of stopping it. Sometimes, therapists asked patients to tape-record their own thoughts about their behavior and why it was good or enjoyable, and then that audiotape was overlaid with messages repeatedly using the words "sickening" and "nauseating," and describing how the person needed to give it up to live a so-called normal life. In one 1962 case, a male homosexual with a Kinsey 6 rating (attraction to his own sex alone) was subjected to the juxtaposition of a tape-recording of his verbal fantasies with nausea drugs every two hours for thirty hours straight. After a twenty-four-hour rest, the treatment was repeated for another straight thirty-two hours (James 1962). Thus, Aldous Huxley's *Brave New World,* with its depiction of sleep-based mental-conditioning audio for children, came to life for adults (Huxley 1932). The effect of using a patient's own words and interests against them was depicted clearly in *A Clockwork Orange.* Some patients, who were deprived of sleep for long periods, felt temporarily cured, but some also experienced complete nervous breakdowns, and one developed a heart condition as a result.

Psychiatrists in the late 1950s and early 1960s, many of whom invested their careers in the idea of medicine as a therapeutic vehicle to normalcy, began to turn their attention to the usefulness of electric shock therapy as a replacement for nausea drugs. Descriptions of the shock treatment often sound like this 1964 article by R. J. McGuire titled "Aversion Therapy by Electric Shock: A Simple Technique":

> The use of the [electroshock] apparatus follows classical conditioning technique. The stimulus to which aversion is to be produced is presented, often by having the patient imagine the stimulus. This procedure is repeated throughout the treatment session of 20 to 30 min, which can be held from six times per day to once a fortnight. The strength of the shock should be adjusted so it is as painful as the patient can bear... After initial instruction he can treat himself and may take the apparatus home to continue the treatment there (McGuire and Vallance 1964, p. 151).

Such an arrangement also "saves the therapist's time" (ibid.). The advantages of electroshock for changing sexual desire were frequent subjects in medical literature; other authors wrote that "electric aversion has largely supplanted chemical aversion as it is safer, easier to control, more precisely applied, and less unpleasant" (Bancroft and Marks 1968, p. 30). The literature on electroshock aversion therapy frequently compared it to aversion therapy using nausea drugs, claiming how much easier and less messy the former was to administer (William 2010). Another advantage, according to one therapist, included that the "therapist is in a position to administer a discrete stimulus of precise intensity for a precise duration of time at precisely the required moment" (Rachman 1965, p. 291). Thus "the greater control which is possible with electrical stimulation should ensure more effective treatment, closer definition of the treatment process and increased theoretical clarity" (ibid.).

The electroshocks were administered with a cuff attached to a man's arm, wrist, ankle, or leg, or via an electrified grid on the floor on which the patient sat or stood

---

[3] A female doctor remembered how machines separated patients and medical staff during treatments: "It was the nurse therapists who actually gave the aversion therapy, although it was automated, and the nurse therapist would sit in another room when the treatment was taking place. I can't remember now whether they had a one-way mirror or something like that... There

with bare feet (McGuire and Vallance 1964; Audrey 2010). Sometimes the shocks happened while the patient observed films or slide photographs of nude or semi-nude males (known or unknown to him); sometimes they were administered at random; and occasionally at other times when the patient signaled the therapist that his homosexual fantasy was clear.[3] The shock treatments were administered for four to eight weeks at a time, several days per week, on-and-off for ten- to thirty-minute periods. One psychiatrist wrote that the treatments "continued until either a change of interest occurs or it becomes clear that no change is likely. A number of patients...have discontinued treatment of their own accord" (Mac-Cullough and Feldman 1967a, p. 594). The practitioners sometimes shocked themselves to test the machines and sometimes by accident, as one psychologist active in the practice from 1966 through 1971 stated: "It was most unpleasant and the voltage was a jagged alternating voltage so that it had a very unpleasant fizzing feel. I had hundreds of shocks from the thing—sometimes not intended—but we did test it very thoroughly on ourselves and we all knew exactly what it was like" (James 2010, p. 3; see also Paul 2010).

The electroshock machines themselves were often cheaply and poorly designed. A male nurse remembered: "The treatments were so contrived! I mean to see a doctor coming in with a slide projector and a handful of slides, and setting it up, and then putting a couple of electrodes on this lad's body, and plugging him [in]to this machine—it was even crueler than ECT... And I remember asking the Charge Nurse: 'By administering the shock where is the treatment?' And of course this was regarded as an insolent and impertinent question" (Dickinson 2012, p. 304). A man who was an eighteen-year-old apprentice electrician at a hospital in the early 1960s described in an interview that a psychiatrist asked him to build a machine that would shock patients according to when the patient switched one or two 16 mm film projectors (one showing male images, one showing female images) on or off. The doctor had no money for the machine, so the electrician built it using spare parts. He built a contraption with a copper tube that measured shaking on the bed, and a leather strap connected the patient's penis to a wire, which then linked to the shock machine. He did not seem to have much faith that the machine worked, or that the patients wanted to be cured: "The patients did it to please [the

---

(Footnote 3 continued)
was some sort of physical barrier between the nurse and the patient" (Elliot 2010, p. 2). A junior male psychologist recalled how he could sometimes not even see the patients whom he was treating: "You didn't really have that much contact with the patients. The room was basically a broom cupboard. Literally it had been a broom cupboard and it was cleared out or so I was told.... I would strap two electrodes on the wrist. Occasionally I seem to remember some people seemed to prefer it on the leg. But it was mostly on the wrist. They'd be on one side of the screen where you projected the slides and I'd be on the other side. You really wouldn't see them even if the light was on. I'd be on the other side operating the projector and the electric shock apparatus" (Isaac 2010, pp. 1–2).

[4] "Heath Robinson" is a term used in British English, named for the cartoonist W. Heath Robinson (1872–1944), which refers to a machine or device that is unnecessarily complicated to achieve a simple task (Carswell 2012).

doctor] so they wouldn't have to go to prison" (George 2010, pp. 2–3). A man in his first year of residency as a clinical psychologist was assigned to administer shock treatments: "It would be the patient and I in a darkened room with an elaborate Heath Robinson set up that the psychiatrist had built.[4] The psychiatrist had practical skills of hydraulics and electric switches and he just happened to put it together. It was very amateurish" (Isaac 2010, p. 1; see also Audrey 2010). The crudeness of these machines showed how under-theorized and under-tested they were. The doctors themselves unsure of what treatments were effective, so they tested a variety of electroshock machines at patients' expense.

One homosexual participant in a 2004 interview series with British aversion therapy patients was coerced into receiving treatment when an undercover police officer entrapped and arrested him in a public place for importuning: "I was given the option, prison or hospital... [I] knew it was not going to make me straight, I didn't want it to, but it seemed a better option than prison" (Smith et al. 2004, p. 428). In the same interview series, a cross-dressing patient recalled how electroshock therapy was conducted in a British National Health Service hospital: 'I remember sitting in the room on a wooden chair "dressed" [in women's clothes], but I had to be barefoot as my feet had to touch the metal electric grid. My penis was also wired up to something to measure if I got an erection—I felt totally violated... I remember the excruciating pain of the initial shock; nothing could have prepared me for it. Tears began running down my face and the nurse said: "What are you crying for? We have only just started!"' (ibid.). All participants in the interview series (seven male, two now female) reported that the treatment they received still disturbed them. Said one, 'I can still have terrible flashbacks of my time in hospital and the barbaric treatments I received.' Said another, 'I just don't know how something so tortuous could have been concealed under the term "health care"' (ibid).

In 1966, the British physician John Bancroft and two colleagues decided to put electroshock therapy together with the penile strain gauge (Bancroft et al. 1966). Combining the two machines was a way to harness the power of them both at the same time: the strain gauge detected erections, and then the shock device would automatically shock the patient if his erection went over a certain circumference. Previous electroshock treatment depended on the therapist or an assistant to administer the shocks, and on the patient who reported when he was about to have an orgasm, when a cross-dresser was wearing women's clothing, or when a gay man or cross-dresser was having a sexual fantasy. Joining these devices distanced therapists further from what they were doing to the patient. They did not have to notice when the patient had an erection or have to listen to them speak about or notice their desires; the machines together did the work for them (Fig. 2.2).

Machine-based aversion therapy was an attempt to do two things: to rid the individual of homosexual feelings and desires and to replace them with heterosexual feelings and desires. Most of the time, however, it erased homosexual feelings and desires by replacing them with no sexual desires at all. Such negative effects from the treatment were fairly common. In a feasibility study of ten men whom Bancroft treated in 1969, one developed phobic anxiety to attractive men

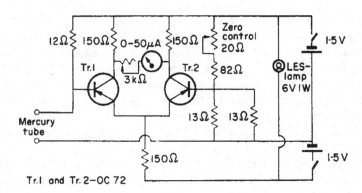

Fig. 2.2 A circuit diagram of a penile strain gauge transducer. Bancroft et al. (1966), p. 239, Fig. 2. Copyright Elsevier, 1966

and attempted suicide; one became aggressive, attempted suicide and was anorgasmic in homosexual relationships; one developed serious rejection by women; one became psychotically depressed and wandered into the streets removing his clothes, and one became disillusioned by the homosexual world and could no longer sustain emotionally rewarding relationships (Bancroft 1969; Bancroft 1974; Smith et al. 2004). On the other hand, more machines did not equal better results, and doctors frequently expressed their frustration with their lack of permanent and lasting success. The treatments seemed to turn patients straight in the short-term—three to six months—if they had some heterosexual experience or interests before they began therapy. But as more time passed, the effects of the aversion therapy would wear off, and for "Kinsey 6s," or men and women with no heterosexual inclinations at all, the treatment had little to no effect on their homosexual desire or orientation (MacCullough and Feldman 1967a, p. 594; James and Early 1963).

Doctor's descriptions of aversion therapy in this era echo Jürgen Habermas's description of the dangers of allowing the values of technology producers and users, specifically efficiency and productivity, to direct workflow in industrial societies rather than allowing for debates over such norms to occur (Habermas 1970; Bimber 1994). The use of aversion therapy echoed contemporary Western values of the 1960s that advocated the incorporation of ever newer, better, and faster machines and technology into everyday private and public life. Electroshock aversion therapy may have looked tidier than other therapies from the outside, but its lack of effectiveness and the deep damage it caused patients left mental scars that haunt them into the present.

Resistance

Psychiatrists and other physicians did not universally accept aversion therapy as a treatment for homosexuality, cross-dressing, and fetishism. After reading the

MacCullough and Feldman study, a physician wrote a letter to the *British Medical Journal* criticizing the premise of the study itself, that homosexuals should adjust to societal conventions instead of the reverse: "because of these persons' unconventional sexual practices they have been conditioned to conventional ones, so much so that it looks as though fornication has become an end in itself" (Kalcev 1967, p. 436; MacCullough and Feldman 1967a). Kalcev extended the authors' thinking to others who did not conform to contemporary standards of social behavior: "If this treatment gives such satisfactory results is there any reason why enthusiasts should not extend it to conditions such as promiscuity or adultery? And if we extend it to people with less conventional views than ourselves in general we will be in 1984" (Kalcev 1967, p. 436). Kalcev's reference to George Orwell's novel, which depicted a world in which the government intensively scrutinized all citizens' behavior and forcibly purged them of any wrongdoing, struck a nerve with MacCullough and Feldman (Orwell 1949). They countered: "The argument that it is society that is wrong is very misguided. Which seems more unethical: to treat someone in distress, or to suggest to him that he waits until his practice is as socially acceptable as heterosexuality?" (MacCullough and Feldman 1967b). Other possibilities for change, such as homosexuals and their allies working to change society to improve tolerance for people of diverse sexualities, did not occur to them.

While aversion therapy caused many individuals long-lasting mental pain, physical pain and distress, some fought the treatment in ways that physicians did not expect. One reviewer noted some resistance even to basic sex-related experiments that doctors conducted with machines, let alone aversion therapy. Marvin Zuckerman wrote in a 1971 review article, "confronted with prying experimenters attaching electrodes, penile plethysmographs, vaginal devices, and showing pornographic stimuli, many subjects might be inclined to inhibit voluntary response" (Zuckerman 1971, p. 326). Subjects of voluntary sex research often found the machines used to play them images, films, or audiotapes, and to measure their responses strange or invasive, and consciously or unconsciously limited their responses as a reaction to machine-based interactions. So too did the subjects of machine-based aversion therapy for non-heteronormative sexualities. Some patients challenged the machines that hospital staffs used to try and change them. Sometimes, with the help of sympathetic nurses who were themselves opposed to aversion therapy, they would work the system in order to bring their treatment to an end.

For example, physicians were unprepared for the possibility that patients might find the electroshocks themselves erotic. Bancroft and Marks noted in 1968 that two of four male sadomasochists receiving electroshock treatment to arrest their sadomasochism had an increase, rather than a decrease, in their erections during the treatment (Bancroft and Marks 1968, p. 32). Despite some sadomasochists revealing their pleasure in the application of pain, limiting its usefulness as a deterrent, Bancroft and his associates kept trying to use electroshock to rid them of such desire. According to him in 1971, for one patient, "When the masochistic fantasy was involved, the anticipation of a shock resulted in greater erections. As treatment continued there was an increase not only in his erections to masochistic fantasies but also [in] his subjective ratings of sexual interest... It seems possible that the aversive procedure was aggravating the masochistic tendency" (Bancroft 1971, p. 127; see also Sandler 1964). Far from the electrical shock averting sado-masochistic or homosexual feelings, then, the shock became part of the patient's normal sexual fantasies. Thus, for some individuals, the pain of shock caused them pleasure, a situation that the medical staff clearly had no idea how to handle. If pain was incorporated into pleasure, what other kind of punishment for deviant behavior could they administer? The fact that some patients were given small shock boxes for home use would have excited someone who found erotic stimulation in pain (McGuire and Vallance 1964).[5] It is clear from the above examples that patients resisted treatments in unanticipated ways.

Nurses and doctors assigned to aversion therapy wards in the 1960s had mixed emotions about and reactions to participating in aversion treatment procedures. Many were disturbed whenever their colleagues seemed to get a thrill out of giving the shocks to patients. A British nurse interviewed in 2010 remembered that "there were also some nurses who you could tell enjoyed administering these aversion treatments. There were others, myself included, who never enjoyed this aspect of their role and considered it barbaric" (Dickinson 2012, p. 202). A female doctor recalled: "I was asked to show some nursing staff and medical students the electric shock equipment, and I picked it up and held these electrodes in my hand and suddenly pressed the button and it was painful. I thought, how can anyone subject themselves to having this? You questioned whether you were getting into a sadomasochistic role" (Elliot 2010, p. 2). Some practitioners denied that they received any pleasure from administering the treatments. "I can't say I particularly enjoyed administering these treatments. I'm not a sadist at heart" (Isaac 2010, p. 3). When Bancroft reflected on his work in aversion therapy in a 1999 *New York Times* interview, he said that his participation in it embarrassed him now, but that "my motives for doing it were entirely honorable. I just think it was a stage of development in the way we were thinking about it" (Dreifus 1999). His motivation for participating in aversion therapy was to help people that had asked him to help them change, which was rarely a request he heard anymore (ibid.). Physicians who used aversion therapy had a variety of motives for their involvement in it, including a desire to help, a desire for fame or notoriety, a desire to experiment,

---

[5] The home-use shock boxes did not work any better than the laboratory-based shock treatment. The fiancée, later wife of a man who was given a shock box remembered it this way: "It was a small wooden box which gave him electric shocks. It would be about 4 or 5 inches square with a dial on it and wires. He explained that he'd been told that whenever he felt interest in a man he should give himself this treatment. I never witnessed him using it, so I don't know how it worked specifically or what his immediate reaction to it was when he gave himself the treatment. He explained that he did use this from time to time and that he felt that the advice he'd been given was that this would help him to get over these homosexual feelings and that he should be able then to lead a 'normal life.'... For it to work he'd have had to carry it round in his pocket and every time you saw a man you fancied you'd have to go ahead and give yourself an electric shock in order to make sure you took action on any possible opportunity. He'd been assured that by giving himself this treatment this would suppress the urges and eventually they would go away" (Joan 2010, p. 1). See also (Tony 2010).

and perhaps a desire to harm. Whatever those motivations were, the therapy did the majority of patients more harm than good.

Subordinate medical staff with direct contact to patients saw their suffering and tried to alleviate it. One nurse recalled that she only pretended to give a patient a required injection: "It sickened me knowing what we had to do to him in the futile hope of making him heterosexual. I just thought: 'Where is the treatment in that? I just couldn't see any benefit to it—it was punishment and torture'" (Dickinson 2012, p. 242). After a conversation with the patient, she faked giving him the injections, and told him to lie to other staff members that he felt the treatment working and that he was feeling more attracted to women. Her coaching worked, and the man was discharged a few weeks later. He sent her a thank-you note, and told her that he was now living with his boyfriend from before his hospitalization. Both a nurse and a patient at the same hospital remember an incident in which the patient had to receive "social skills training" by going out on a pretend date on the hospital grounds (ibid., pp. 244, 248). Instead the patient did humorous impressions of hospital staff members, the nurse lied to the matron about her impressions of the patient's progress toward becoming heterosexual, and the patient was discharged a few weeks later.

A male ex-patient reflected around forty years after his treatment in Britain on his experiences with physicians administering aversion therapy:

> I don't think they were either driven or cruel sadist[s]. I think they were deluded by a body of theory, particularly in the institution they worked in, that suggested that this was a successful way of dealing with people with obsessive behaviours as they probably see

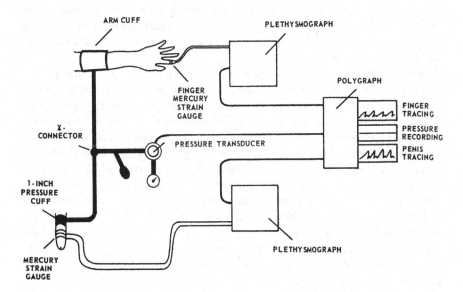

Fig. 2.3 Schematic arrangement of mercury strain gauge plethysmography used to estimate penile blood flow to indicate impotence. Britt et al. (1971) p. 674, Fig. 1. Copyright Elsevier, 1971

them. Money is much better spent counseling people to come to terms with their sexuality and live a wholesome and full life and to learn to love themselves again—not inflict more harm (Warren 2010, p. 4).

Unlike physicians, who wanted to adjust patients to the widely anti-homosexual social attitudes and mores at the time, this former patient instead thought that society should let people of differing sexualities live fulfilling and happy lives. Rather than teaching self-hate, this man argued, medicine should promote self-acceptance. In the present, this former patient's view of how Anglophone society should operate has been more prevalent, though the idea that homosexuality is a disease has remained (Fig. 2.3).

Conclusion

Aversion therapy slowly lost favor in the Anglophone medical community in the late 1960s and early 1970s as gay and lesbian movements, among other civil rights movements, gained strength. The Sexual Offences Act of 1967 partially decriminalized homosexuality in England and Wales. The Stonewall riots in New York City, which occurred in June 1969 when gay, lesbian, and transgender patrons of the Stonewall Inn fought back against a police raid, galvanized the American homosexual rights movement (Duberman 1994). Two students at the London School of Economics, inspired by the events at Stonewall, created the Gay Liberation Front (GLF) in October 1970 to form a united front against homosexual discrimination (Cook 2007). The American Psychological Association (APA) removed homosexuality from its classification of mental disorders in 1973. The medical, legal, and cultural barriers to homosexuals being allowed full civil rights were beginning to crumble.

This analysis of the use of the penile strain gauge, electroshock therapy, and associated technologies illustrates a recurring theme in the history of technology and its intersection of the history of gender and sexuality: as Nina E. Lerman, Ruth Oldenziel, and Arwen Mohun put it in the introduction to *Gender and Technology,* "When we recognize the role of human choice in shaping social and technical categories and their relationships, we must recognize also that both gender and technology are about power: social, cultural, economic, political" (Lerman et al. 2003, p. 7). Users determine the values of technological devices, and cultural, medical, and legal systems empower those individuals to use devices on or with particular individuals. A political and cultural preference for containment authorized the medical establishment to use technology to cause pain to sexual deviants in the name of disciplining their bodies, curing them, and making them "normal." But professional organizations, government bodies, or the patients themselves could also take away that power.

Moreover, the broader history of sex research devices shows that individuals are not merely passive victims of a techno-medical agenda. They also challenge such agendas, and even unorganized and unrecorded resistances matter for eventual shifts in attitudes and values. For example, the behavior of the sadomasochists in Bancroft's article exemplifies the idea of everyday resistance. As James Scott wrote in his now-classic text *Weapons of the Weak*, "Where everyday resistance most strikingly departs from other forms of resistance is in its implicit disavowal of public and symbolic goals" (Scott 1985, p. 31). Additionally, Scott argues that "when [individual] acts are rare and isolated, they are of little interest; but when they become a consistent pattern (even though uncoordinated, let alone organized) we are dealing with resistance" (ibid., 296). Thus some individuals subject to aversion therapy were able to resist it, and to use its own tools and machines against itself.

Nussbaum's theoretical approach sheds light on the fact that men subject to aversion therapy were treated more as objects than as subjects. Anglophone culture at large considered their sexual desires and behaviors so distasteful that it allowed the medical establishment to use machines to deny their subjectivity and agency. Aversion therapy patients were treated more as fungible things, with no feelings or emotions to be accounted for, than as human beings with autonomy. As the doctors sat far away from Alex and the movie projector in *A Clockwork Orange,* so too did the machines of aversion therapy in real life permit doctors to treat those patients with significant distance from their own feelings.

Voluntary human sex research in the 1950s and 1960s was also limited in what it could find on account of the rooms, instruments, and close involvement of the researchers themselves. Subjects could not often freely move due to being hooked up to machines; they may have been shy or intimidated by the immediate presence of the technicians, or found the still images, audiotapes, or films un-stimulating. Researchers usually presumed that their subjects were heterosexual, and they were unsure what to think or how to respond when study subjects behaved differently than expected, particularly when subjects had responses to same-sex images. However, researchers using voluntary subjects were relatively open-minded regarding variation in their subjects' possible responses. Further, study subjects, as noted above, could disrupt observations, intentionally or not, or could reject researchers' findings to make their own meanings out of the data that they helped create. Sex research, then, was breaking out of the laboratory spaces in which it was confined and moving out into the wider world.

References

Aruta, A. 2011. Shocking waves at the museum: The Bini-Cerletti electro-shock apparatus. *Medical History* 55(3): 407–412.

Audrey. 2010. Using desensitisation. http://www.treatmentshomosexuality.org.uk/images/documents/professional/Using_desensitisation.pdf. Accessed 20 Mar 2013.

Bancroft, J. 1969. Aversion therapy of homosexuality: A pilot study of 10 cases. *The British Journal of Psychiatry* 115(12): 1417–1431. doi:10.1192/bjp.115.529.1417.

Bancroft, J. 1971. The application of psychophysiological measures to the assessment and modification of sexual behaviour. *Behaviour Research and Therapy* 9(2): 119–130. doi:10.1016/0005-7967(71)90069-6.

Bancroft, J. 1974. *Deviant sexual behaviour: Modification and assessment.* Oxford: Oxford University Press.

Bancroft, J. 2007. General discussion. In *The psychophysiology of sex*, ed. E. Janssen. Bloomington: Indiana University Press.

Bancroft, J., and I. Marks. 1968. Electric aversion therapy of sexual deviations. *Proceedings of the Royal Society of Medicine* 61(8): 796–799.

Bancroft, J.H.J., H.G. Jones, and B.R. Pullan. 1966. A simple transducer for measuring penile erection, with comments on its use in the treatment of sexual disorders. *Behaviour Research and Therapy* 4(3): 239–242. doi:10.1016/0005-7967(66)90075-1.

Barlow, D.H., et al. 1970. A mechanical strain gauge for recording penile circumference change. *Journal of Applied Behavior Analysis* 3(1): 73–76. doi:10.1901/jaba.1970.3-73.

Bell, A.I. 1975. Male anxiety during sleep. *International Journal of Psychoanalysis* 56: 455–464.

Bell, A.I., and C.F. Stroebel. 1973. The scrotal sac and testes during sleep: Physiological correlates and mental content. In *Sleep: Physiology, biochemistry, psychology, pharmacology, clinical implications*, ed. W.P. Koella, and P. Levin, 380–384. Basel: S. Karger.

Bernick, N., A. Kling, and G. Borowitz. 1971. Physiologic differentiation of sexual arousal and anxiety. *Psychosomatic Medicine* 33(4): 341–352.

Bimber, B. 1994. Three faces of technological determinism. In *Does technology drive history?*, ed. M.R. Smith, and L. Marx, 80–100. Cambridge: MIT Press.

Britt, D.B., W.T. Kemmerer, and J.R. Robison. 1971. Penile blood flow determination by mercury strain gauge plethysmography. *Investigative Urology* 8(6): 673–678.

Brown, C.G. 2011. Sex, religion, and the single woman c. 1950–75: The importance of a "short" sexual revolution to the English religious crisis of the sixties. *Twentieth-Century British History* 22(2): 189–216. doi:10.1093/tcbh/hwq048.

Bryant, C. 2012. Turing centenary: The trial of Alan Turing for homosexuality. http://www.polarimagazine.com/features/turing-centenary-trial-alan-turing-homosexual-conduct. Accessed 17 Mar 2013.

Carswell, B. 2012. W. Heath Robinson: A complex machine. http://www.abebooks.com/books/illustration-art-uncle-lubin-william/w-heath-robinson.shtml. Accessed 21 Mar 2013.

Chivers, M.L., and J.M. Bailey. 2007. The sexual psychophysiology of sexual orientation. In *The psychophysiology of sex*, ed. E. Janssen, 458–474. Bloomington: Indiana University Press.

Cohen, D. 2013. *Family secrets: Living with shame from the Victorians to the present day.* London: Viking.

Cook, M. 2007. Queer conflicts: Love, sex and war, 1914–1967. In *A gay history of Britain: Love and sex between men since the middle ages*, ed. M. Cook, 145–178. Oxford: Greenwood World Publishing.

Cuordileone, K.A. 2000. "Politics in an age of anxiety": Cold war political culture and the crisis in American masculinity, 1949–1960. *Journal of American History* 87(2): 515–545.

Cuordileone, K.A. 2005. *Manhood and American political culture in the Cold War.* New York: Routledge.

Daily Mail. 2007. Lord Montagu on the court case which ended the legal persecution of homosexuals. http://www.dailymail.co.uk/news/article-468385/Lord-Montagu-court-case-ended-legal-persecution-homosexuals.html#. Accessed 19 Mar 2013.

D'Emilio, J., and E. Freedman. 1988. *Intimate matters: A history of sexuality in America.* New York: Harper & Row.

Davis, R.L. 2010. *More perfect unions: The American search for marital bliss.* Cambridge: Harvard University Press.

Dickinson, T. 2012. *Mental nursing and "sexual deviation": Exploring the role of nurses and the experience of patients, 1935–1974.* Dissertation: University of Manchester.

Dreifus, C. 1999. A conversation with John Bancroft: Sitting in the ultimate hot seat; the Kinsey Institute. New York Times. http://www.nytimes.com/1999/05/25/science/conversation-with-john-bancroft-sitting-ultimate-hot-seat-kinsey-institute.html. Accessed 21 Mar 2013.

Drucker, D.J. 2010. Male sexuality and Alfred Kinsey's 0–6 scale: Toward "a sound understanding of the realities of sex". *Journal of Homosexuality* 57(9): 1105–1123. doi:10.1080/00918369.2010.508314.

Duberman, M. 1994. *Stonewall*. New York: Plume.

Elliot. 2010. Doubts about the treatment. http://www.treatmentshomosexuality.org.uk/images/documents/professional/Doubts_about_the_treatment.pdf. Accessed 20 Mar 2013.

Farkas, G.M., et al. 1979. Reliability and validity of the mercury-in-rubber strain gauge measure of penile circumference. *Behavior Therapy* 10(4): 555–561. doi:10.1016/S0005-7894(79)80056-8.

Foucault, M. 1977. *Discipline and punish: The birth of the prison*. Trans: A. Sheridan. New York: Vintage.

Freund, K. 1957. Diagnostika homosexuality u muza [Diagnosing homosexuality in men]. *Cs Psychiatry* 53: 382–393.

Freund, K. 1963. A laboratory method for diagnosing predominance of homo- or hetero-erotic interest in the male. *Behaviour Research and Therapy* 1(5): 85–93. doi:10.1016/0005-7967(63)90012-3.

Freund, K. 1965. Diagnosing heterosexual pedophilia by means of a test for sexual interest. *Behaviour Research and Therapy* 3: 229–234. doi:10.1016/0005-7967(65)90031-8.

Freund, K. 1991. Reflections on the development of the phallometric method of assessing erotic preferences. *Sexual Abuse: A Journal of Research and Treatment* 4(3–4): 221–228. doi:10.1007/BF00850054.

Freund, K., J. Diamant, and V. Pinkava. 1958. On the validity and reliability of the phalloplethysmographic (Php) diagnosis of some sexual deviations. *Review of Czechoslovak medicine* 4: 145–151.

Freund, K., F. Sedlacek, and K. Knob. 1965. A simple transducer for mechanical plethysmography of the male genital. *Journal of the Experimental Analysis of Behavior* 8: 170–171. doi:10.1901/jeab.1965.8-169.

Gathorne-Hardy, J. 1998/2004. *Sex the measure of all things: A life of Alfred C. Kinsey*. Bloomington: Indiana University Press.

George. 2010. An apprentice electrician. http://www.treatmentshomosexuality.org.uk/images/documents/professional/An_apprentice_electrician.pdf. Accessed 20 Mar 2013.

Gibbons, G.E., D.E. Strandness Jr, and J.W. Bell. 1963. Improvements in design of the mercury strain gauge plethysmograph. *Surgery, Gynecology, and Obstetrics* 116: 679–682.

Grosz, E. 1995. Bodies and knowledges: Feminism and the crisis of reason. In *Space, time, and perversion: Essays on the politics of bodies*, 25–43. New York: Routledge.

Habermas, J. 1970. *Towards a rational society*. Boston: Beacon.

Hilgard, E.R. 1961. *Hilgard and Marquis' conditioning and learning*. 2nd ed. Kimble GA (rev), New York: Appleton-Century-Crofts.

Huxley, A. 1932. *Brave new world*. London: Chatto & Windus.

Isaac. 2010. Relief and shame. http://www.treatmentshomosexuality.org.uk/images/documents/professional/Relief_and_shame.pdf. Accessed 27 Apr 2013.

James. 2010. How views change. http://www.treatmentshomosexuality.org.uk/images/documents/professional/How_views_change.pdf. Accessed 20 Mar 2013.

James, B. 1962. Case of homosexuality treated by aversion therapy. *British Medical Journal* 5280: 768–770.

James, B., and D.F. Early. 1963. Aversion therapy for homosexuality. *British Medical Journal* 5329: 538.

Joan. 2010. The wife's story. http://www.treatmentshomosexuality.org.uk/images/documents/patient/The_wife_s_story.pdf. Accessed 20 Mar 2013.

Johnson, D.K. 2004. *The lavender scare: The cold war persecution of gays and lesbians in the government*. Chicago: University of Chicago Press.

Kalcev, B. 1967. Aversion therapy of homosexuals. *British Medical Journal* 5562: 436.

Kinsey, A.C., W.B. Pomeroy, and C.E. Martin. 1948. *Sexual behavior in the human male.* Philadelphia: W. B. Saunders.

Latour, B., and S. Woolgar. 1986. *Laboratory life: The construction of scientific facts.* Princeton: Princeton University Press.

Laws, D.R. 1977. A comparison of the measurement characteristics of two circumferential penile transducers. *Archives of Sexual Behavior* 6(1): 45–51. doi:10.1007/BF01579247.

Laws, D.R., and R.A. Bow. 1976. An improved mechanical strain gauge for recording penile circumference change. *Psychophysiology* 13(6): 596–599. doi:10.1111/j.1469-8986.1976.tb00888.x.

Laws, D.R., and A.V. Pawlowski. 1973. A multi-purpose biofeedback device for penile plethysmography. *Journal of Behavior Therapy and Experimental Psychiatry* 4(4): 339–341. doi:10.1016/0005-7916(73)90004-9.

Laws, D.R., and H.B. Rubin. 1969. Instructional control of an autonomic sexual response. *Journal of Applied Behavior Analysis* 2(2): 93–99. doi:10.1901/jaba.1969.2-93.

Lerman, N.E., R. Oldenziel, and A. Mohun. 2003. *Gender and technology: A reader.* Baltimore: Johns Hopkins University Press.

Lewis, C.H. 2010. *Prescription for heterosexuality: Sexual citizenship in the cold war era.* Chapel Hill: University of North Carolina Press.

Liversedge, L.A., and J.D. Sylvester. 1955. Conditioning techniques in the treatment of writer's cramp. *Lancet* 1: 1147–1149.

Loiselle, R.H., and S. Mollenauer. 1965. Galvanic skin response to sexual stimuli in a female population. *The Journal of General Psychology* 73(2): 273–278. doi:10.1080/00221309.1965.9710721.

MacCullough, M.J., and M.P. Feldman. 1967a. Aversion therapy in management of 43 homosexuals. *British Medical Journal* 5552: 594–597.

MacCullough, M.J., and M.P. Feldman. 1967b. Aversion therapy in management of homosexuals. *British Medical Journal* 5570: 51–52.

Max, L.M. 1935. Breaking up a homosexual fixation by the conditioned reaction technique: A case study. *Psychological Bulletin* 32(11): 734.

May, E.T. 1988. *Homeward bound: American families in the cold war era.* New York: Routledge.

McConaghy, N. 1967. Penile volume change to moving pictures of male and female nudes in heterosexual and homosexual males. *Behaviour Research and Therapy* 5(1): 43–48. doi:10.1016/0005-7967(67)90054-X.

McGuire, R.J., and M. Vallance. 1964. Aversion therapy by electric shock: A simple technique. *British Medical Journal* 5376: 151–153.

McLaren, A. 2007. *Impotence: A cultural history.* Chicago: University of Chicago Press.

Minton, H.L. 1988. *Lewis M. Terman: Pioneer in psychological testing.* New York: New York University Press.

Mort, F. 2010. *Capital affairs: London and the making of the permissive society.* New Haven: Yale University Press.

Nussbaum, M.C. 1995. Objectification. *Philosophy & Public Affairs* 24(4): 249–291. doi:10.1111/j.1088-4963.1995.tb00032.x.

Ohlmeyer, P., and H. Brilmayer. 1947. Periodische vorgange im schlaf [Periodic events in sleep]. *Pflügers Archiv für die gesamte Physiologie des Menschen und der Tiere* 248: 559–560. doi:10.1007/BF00362669.

Ohlmeyer, P., H. Brilmayer, and H. Hullstrung. 1944. Periodische vorgange im schlaf II [Periodic events in sleep II]. *Pflügers Archiv für die gesamte Physiologie des Menschen und der Tiere* 249: 50–55. doi:10.1007/BF01764449.

Orwell, G. 1949. *1984.* New York: New American Library.

Paul. 2010. Unease about the treatment. http://www.treatmentshomosexuality.org.uk/images/documents/professional/Unease_about_the_treatment.pdf. Accessed 20 Mar 2013.

Perry, C.C., and H.R. Lissner. 1955. *The strain gauge primer.* New York: McGraw-Hill.

Rachman, S. 1961. Sexual disorders and behavior therapy. *The American Journal of Psychiatry* 118(3): 235–240.

Rachman, S. 1965. Aversion therapy: Chemical or electrical? *Behaviour Research and Therapy* 2: 289–299. doi:10.1016/0005-7967(64)90036-1.

Ramsey, G.V. 1943. The sex information of younger boys. *American Journal of Orthopsychiatry* 13(2): 347–352. doi:10.1111/j.1939-0025.1943.tb06004.x.

Raymond, M.J. 1956. Case of fetishism treated by aversion therapy. *British Medical Journal* 4997: 854–857.

Raymond, M., and K. O'Keeffe. 1965. A case of pin-up fetishism treated by aversion conditioning. *The British Journal of Psychiatry* 111(7): 579–581. doi:10.1192/bjp.111.476.579.

Reumann, M.G. 2005. *American sexual character: Sex, gender, and national identity in the Kinsey reports.* Berkeley: University of California Press.

Rosen, R.C. 1973. Suppression of penile tumescence by instrumental conditioning. *Psychosomatic Medicine* 35(6): 509–514.

Rosen, R.C., and F.J. Keefe. 1978. The measurement of human penile tumescence. *Psychophysiology* 15(4): 366–376. doi:10.1111/j.1469-8986.1978.tb01393.x.

Rosen, R.C., and S.A. Kopel. 1977. Penile plethysmography and biofeedback in the treatment of a transvestite exhibitionist. *Journal of Consulting and Clinical Psychology* 45(5): 908–916. doi:10.1037/0022-006X.45.5.908.

Rosen, R.C., D. Shapiro, and G.E. Schwartz. 1975. Voluntary control of penile tumescence. *Psychosomatic Medicine* 37(6): 479–483.

Russell, M.A. 1970. Effect of electric aversion on cigarette smoking. *British Medical Journal* 5688: 82–86.

Sandler, J. 1964. Masochism: An empirical analysis. *Psychological Bulletin* 62(3): 197–204. doi:10.1037/h0040597.

Scott, J. 1985. *Weapons of the weak: Everyday forms of peasant resistance.* New Haven: Yale University Press.

Seraine, L. 1865. *De la santé des gens mariés [On the health of married people].* Paris: F. Savy.

Shahid, S. 2008. The Daily News's front-page photo of Ruth Snyder's execution (New York Daily News). http://www.newseum.org/news/2008/01/the-daily-newss-front-page-photo-of-ruth-snyders-execution-new-york-daily-news.html. Accessed 13 Mar 2013.

Smith, G., A. Bartlett, and M. King. 2004. Treatment of homosexuality in Britain since the 1950s—an oral history: the experience of patients. *British Medical Journal* 328(7437): 427–429. doi:10.1136/bmj.37984.442419.EE.

Tony. 2010. Traditional ideas of sex and sexuality. http://www.treatmentshomosexuality.org.uk/images/documents/patient/Traditional_ideas_of_gender_and_sexuality.pdf. Accessed 20 Mar 2013.

Warren. 2010. Family pressures. http://www.treatmentshomosexuality.org.uk/images/documents/patient/Family_pressures.pdf. Accessed 20 Mar 2013.

Whitney, R.J. 1949. The measurement of changes in human limb-volume by means of a mercury-in-rubber strain gauge. *Journal of Physiology* 109(Suppl): 5P–6P.

William. 2010. When the psychiatrist is gay. http://www.treatmentshomosexuality.org.uk/images/documents/professional/When_the_psychiatrist_is_gay.pdf. Accessed 20 Mar 2013.

Wolfenden, J., et al. 1957. *Report of the departmental committee on homosexual offences and prostitution.* London: Her Majesty's Stationery Office.

Wood, D.M., and P.A. Obrist. 1968. Minimal and maximal sensory intake and exercise as unconditioned stimuli in human heart-rate conditioning. *Journal of Experimental Psychology* 76((2, pt. 1)): 254–262. doi:10.1037/h0025373.

Zuckerman, M. 1971. Physiological measures of sexual arousal in the human. *Psychological Bulletin* 75(5): 297–329. doi:10.1037/h0030923.

Chapter 3
The Couples Laboratory and the Penis-Camera: Seeking the Source of Orgasm

Abstract The laboratory, machines, and research of William H. Masters and Virginia E. Johnson in St. Louis, Missouri, are discussed. Masters, an obstetrician and gynecologist at Washington University specializing in infertility, began a research project on sexual physiology in the 1950s as a means to understanding better the relationship between sexual behavior and fertility. With the assistance of Johnson and laboratory technicians, Masters conducted research into the anatomy and physiology of human sexual response by using multiple machines in, on, and around the bodies of single subjects or couples. Those machines included a penis-camera, which filmed inside a woman's vagina as she masturbated with it. Masters and Johnson published their results in academic journals before publishing a compilation of them in *Human Sexual Response* (1966). Their best-known finding was the four-stage response cycle, outlining the four steps that women's and men's bodies went through in a sexual experience: excitement, plateau, orgasm, and resolution. The main difference they found between men and women was that women could have multiple orgasms without resolution, and men could not. Masters and Johnson used their machine-based results as evidence to argue that heterosexual married couples could aim for parity in their sexual gratification. However, homosexual men and women, along with radical feminists, read their work as evidence for sexual liberation and satisfaction outside of marriage as well. More conservative commentators objected that Masters and Johnson's research took the mystery out of sex. *Human Sexual Response* cemented the importance of using machines in sex research.

Keywords Masters and Johnson · Sex research · Human sexuality · Four-stage response cycle · Orgasm · Human Sexual Response

William H. Masters and Virginia E. Johnson, in their book *Human Sexual Response,* diagrammed the intimate anatomy and physiology of more than six hundred men and women during masturbation and sexual intercourse in a sex research laboratory in St. Louis, Missouri (Masters and Johnson 1966). They used multiple technologies, including motion picture film cameras, heart and blood pressure monitors, and a rheostat-powered dildo-shaped camera that could take

D. J. Drucker, *The Machines of Sex Research,*
SpringerBriefs in History of Science and Technology,
DOI: 10.1007/978-94-007-7064-5_3, © The Author(s) 2014

pictures inside a woman's vulva to gather their data. *Human Sexual Response* is perhaps best known for outlining a four-stage orgasmic response cycle in both men and women that highlighted the importance—and for most women, the necessity—of the clitoris for female orgasm and promoted equal sexual enjoyment for male and female heterosexuals (Masters and Johnson 1966). The machines that Masters and Johnson used to measure sexual response in male and female bodies were central to their research processes for capturing and processing sexual data. Those machines thus created and delineated two new types of sexual space: the cinematically mapped interior of women's bodies and heterosexual bedrooms in which men and women now had similar scientifically proven capacities for sexual pleasure.

Other scholars have written about Masters and Johnson's use of technology in works on American history and sociology, and in histories of sexuality, sex research, and the sexual revolution. They have thoroughly critiqued Masters and Johnson's findings, their obfuscating language, their omission of socio-cultural gender inequality, and their ignorance of historical precedents in delineating the stages of the response cycle (Robinson 1976; Tiefer 1995; Maines 1999; Irvine 2005; Morrow 2008). Most scholarship on the pair focuses on two of their principal findings: first, that the clitoris is central to female orgasm, and second, that women are capable of multiple orgasms (D'Emilio and Freedman 1988; Ericksen and Steffen 1999; Reumann 2005; Davidson and Layder 1994; Allyn 2000). However, examining Masters and Johnson's work with a background of spatial theory, the history of technology, and science studies draws attention to how the researchers used machines to shape their investigative processes and subsequent findings. As Vern L. Bullough has pointed out, a key element in the ability of Masters and Johnson to break new ground was their use of technology (Bullough 1994). Studying Masters and Johnson's research processes illuminates the ways that uses of machines shape scientific and cultural ideas about human sexuality and the spaces in which sexual behavior takes place.

Masters and Johnson's work illustrates the place of machines in cementing the scientific authority of sex research and sexology as a discipline. For many critics, their research—filming and recording human bodies in masturbation and intercourse—was voyeurism, not science. Machines and machine-produced data supported their claims to scientific authority, and the idea that sex research was "real science." Masters and Johnson used their research processes and results to establish sex research as an authoritative field of scientific practice. For some contemporary observers, "the notion of scientists studying the most intimate matters with such clinical detachment and high-tech hardware was startling yet comforting. Their research was widely regarded as reassuring evidence of America's scientific advancement" (Allyn 2000, p. 168). Masters and Johnson, building on Alfred Kinsey's observations and mass-collected statistical data in the 1940s and 1950s, were able to promote further "public acceptance of sex in the laboratory" and to draw on the moral authority that science had in mid-twentieth century American society (Irvine 2005, p. 54; Kinsey et al. 1948, 1953).

This chapter focuses on how Masters and Johnson's machines functioned as tools that created new kinds of sexual space. Spatial theory and visual cultural studies of science illuminate the modern development of physical (interior and exterior to the body) and metaphorical spaces that prioritize those spaces' abilities to provide and to facilitate sexual pleasure above other functions and purposes (Penley and Ross 1991; Lefebvre 1991; Cartwright 1995; Fausto-Sterling 1992, 2000; Maines 1999). The concept of sexual space concerns the complex relationships between persons, objects, and behaviors and the ways that those relationships change over time. Using theories, studies, and histories together shows the ways that sex research, sex researchers, technologies, and bodies operate in tandem. They show how technologies and researchers work together to form and reform bodies and homes in a new, modern light. Technology, specifically the machines of sex research and their users, established the socio-spatiality of sexuality, transformed the ways that people viewed their interior and exterior selves as sexual entities, and reformed private homes into places of non-reproductive sexual pleasure.

This chapter outlines the machines that Masters and Johnson used from their earliest published papers in 1959 through 1970, when they dismantled their research laboratory and turned their attention to marital sex therapy. It then moves to what the machines found in the responses of male and female bodies to sexual arousal, and concludes with analysis of popular critiques of their uses of machines for sex research. While some criticisms of technology argued that machines eliminated the mystery of sexual connection between a man and a woman, the machines nonetheless provided information about men's and women's anatomy and physiology that increased possibilities of orgasmic pleasure for both sexes.

This chapter examines how Masters and Johnson's processes and results made machines (both standard medical equipment and specially invented items) essential tools for human sex research and for understanding bodies as gendered and sexualized spaces. Masters and Johnson debunked the idea that women could achieve orgasm from vaginal penetration alone, and used technology to provide clinical data to identify the clitoris as the primary organ for female orgasm. Despite Masters and Johnson's clear valorization of heterosexuality and marriage, their findings helped foster the American sexual revolution of the late 1960s through the creation of new sexual spaces in bodies and in bedrooms with data captured via machine technology.

The Visible Body in the Laboratory

Masters and Johnson, from their earliest collaborations in 1956, used many different kinds of machines to gather data on human sexual behavior. They used a film camera, a penis-camera (described below), tape recorders, a colposcope, an electroencephalogram, an electrocardiogram, a respirator, and a pH meter with electrode assembly, among other devices to measure sexual behavior. The focus in the discussion below is on visual technologies, as in the hands of Masters and Johnson, visual technologies "provide particular ways of accessing the internal body and determin[ing] its depiction; the resulting representations, in turn, fashion our knowledge of the body and set the parameters of its conceptualization" (van Dijck 2005, p. 11). Through imaging technology, Masters, Johnson, and their associates could make "visible parts of the living body that were previously considered to be too interiorized, too minute, or too private to be seen by the researcher's unaided eye" (Cartwright 1995, p. 23). New ways of seeing interiors and exteriors of bodies led to new ways of seeing sexual, orgasmic bodies.

Masters first thought of using film cameras to record human sexual behavior partway through his tenure as an associate professor in obstetrics and gynecology specializing in infertility at Washington University. His interest in sex research dated back to the late 1930s, but his advisor at the University of Rochester School of Medicine and Dentistry, George W. Corner, suggested that he not begin it until he had achieved tenure and had the backing of a university (Bullough 1994). While tending to patients suffering from infertility, he realized how little physicians, including him, knew about the anatomy and physiology of human sexuality and its relationship to fertility. During a sabbatical in 1954, and with the permission of the Washington University chancellor, the St. Louis chief of police, and the city's Roman Catholic archbishop, Masters began interviewing, viewing, and providing free medical exams to the city's male and female prostitutes on the job. While viewing dozens of prostitutes at work from behind curtains or in corners of their work spaces, he began to envision how to use medical equipment to measure and to record the bodily changes he observed (Maier 2009, p. 81). Quite rapidly, Masters shifted his perceptions of the authority he needed to conduct research from local politicians and prelates to his own judgment, which machines aided and supported.

Masters' (and later Johnson's) early research questions, as stated in *Human Sexual Response,* were: "What happens to the human male and female as they respond to effective sexual stimulation? Why do men and women behave as they do when responding to effective stimulation?" (Masters and Johnson 1966, p. 10). They used those research questions to create and to gather the rhetorical and material tools they needed to answer them, including "the development of effective techniques of interrogation, observation, and physical recording. Once these technical tools were at hand," they began work together in earnest in 1956 (ibid.). They sought to film and to record the movements of and bodily changes in individuals and couples during masturbation and intercourse, setting up a private laboratory with a bed and multiple machines. After data collection, they aggregated the data and began to publish their findings in relatively obscure medical journals.

Masters and Johnson's experiments involved mostly heterosexual couples but also single women, with the head investigators observing from behind a one-way mirror and recording the proceedings on audiotape and film. Masters would also enter the room to observe the use of the penis-camera to study women's interior vaginal changes, or to test for vaginal lubrication or pH balance. Female subjects were required to use the penis-camera at specific times, and couples had to separate periodically during intercourse for vaginal pH testing, making an already awkward and unerotic situation even more so. They recruited subjects for their experiments mapping the physiology of orgasm by asking for volunteers to "engage in overt sexual activity in a laboratory environment" (Maier 2009, p. 105). The laboratory, which was in the Maternity Hospital at Washington University, was "a sparse, almost empty room" with a leather-padded chaise lounge fitted with stirrups, baseboard with multiple outlets, and whatever machines were needed, usually including the penis-camera, electrocardiograph, and electroencephalograph (ibid., p. 99). There was no climate control, so the room was often stuffy and hot.

In advance of filming and recording sessions, Masters and Johnson would show prospective research project volunteers, most of whom were medical students, hospital personnel, and doctor's wives recruited from the Washington University medical system, each of the machines in the research laboratory and would describe what each measured.

> "This is the electrocardiograph," Dr. Masters might explain. "With it we record the detailed behavior of the heart during the four phases of sexual response. This information may prove helpful in counseling elderly people and people with heart disease. There is so much that nobody yet knows—but that we may be able to find out with your help." Or Mrs. Johnson might explain the "pH meter" and the glass-coated electrode attached to it. "One of the subjects we are investigating," she would say, "is the relative acidity of the vagina under various conditions, a factor which may prove significant in helping couples who have been trying without success to have a baby. When the tip of this electrode is touched to the wall of the vagina, this meter records the acidity at the point touched" (Brecher and Brecher 1966, pp. 55–56).

An explanation of the machines, their purposes, and the end goals of the research was necessary for even medically savvy volunteers. While the machines would perform their functions with the occasional adjustment from the quiet yet observant research staff, the explanation served to calm volunteers whose every movement, internally and externally, they would record. Volunteers, whom Masters and Johnson required to be orgasmic for participation in their studies, wanted to understand all aspects of what would be measured and how. Being hooked up to machines could affect their ability to perform sexually whether they were alone, with their spouse, or with a partner the staff chose. Though some volunteer subjects were not able to manifest a full sexual response cycle to orgasm at each filming session, they usually attributed their failure to the lack of foreplay, not to the presence of the machines or the laboratory setting (Masters and Johnson 1966; Maier 2009). One of the male participants, who was a regular participant in filming, had no trouble and adjusted his sexual technique with his partners in order that the cameras could capture every move. "He knew just where the electrocardiogram machine could be found, the wires with electrodes, and the placement of the one-way mirror" (Maier 2009, p. 111). His awareness of the machines, even though they made his movements less natural, aided Masters and Johnson's research process.

Participants could wear masks if they wanted to preserve their anonymity with partners (if the partner was unknown to them) or on the film camera if they were alone. The subjects could only move in a limited fashion due to the size of the bed or lounge and the constraints of the machines to which they were attached. One female participant, an obstetrics nurse, told a friend that "It was very quiet, and there was usually two of them having intercourse and Gini Johnson would come in and do readings on medical equipment, being very quiet not to interrupt them" (Maier 2009, p. 105). Two technicians behind movable screens monitored up to five different machines measuring subjects' heart rates, pulse rates, respiration, blood pressure, and overall physiological changes. The photographic technician K. Cramer Lewis, along with William Slater, a Washington University physiology professor who "checked patients' breathing and other vital signs," together "sat discreetly behind movable screens, monitoring their equipment rather than the faces of those before them" (Maier 2009, p. 138).

Color motion picture filming of sexual behavior in the laboratory had advantages over the simple listening and naked-eye observation that Masters had done when watching prostitutes ply their trade. Masters and Johnson did watch their subjects from behind the low-level technology of a one-way mirror. However, researchers employing film, as they did, could analyze occurrences so brief that they could be otherwise missed or misinterpreted; the film could be repeatedly replayed at regular, high, or low speeds; parts of the anatomy or the speed of bodily processes could be measured precisely; and the film could be preserved and compared with films of the same individual over time or with those of others engaging in the same behaviors. Further, films could later be used for instruction in medical schools or during professional medical presentations, which was especially important as Masters and Johnson forbade the live observations by other physicians and medical students of any sexual behavior sessions (Masters and Johnson 1966). The films were important to establish the veracity of Masters and Johnson's findings and to solidify their professional identities as legitimate sex researchers. As Londa Schiebinger argues, "science is not value neutral but emerges from complex cultural matrices," and Masters and Johnson viewed technology as a means to establishing their culturally challenging work as scientifically objective (Schiebinger 2004, p. 74).

In an early sex research article published before Johnson's formal involvement in the project, Masters stated bluntly the effectiveness and necessity of using motion picture films for accurate findings: "The techniques of colored motion-picture photography have provided definitive analysis of the normal intravaginal anatomical reaction to various phases of the female's sexual response cycle" (Masters 1959, p. 301). Using the data from the six women in the study so far, he briefly described four stages of the female human sexual response cycle: excitement, plateau, orgasm, and resolution (discussed below). Film cameras provided views inside women's vaginas that no one had seen before, as Masters used both a standard film camera and a kind of camera that he uniquely developed. He filmed couples in coitus and women in masturbation using a camera installed inside a clear plastic dildo, called a penis-camera. "The direct observation of the development of vaginal lubrication has been one of the most interesting aspects of the gross anatomical study of the human female's sexual response cycle," he wrote, though in this early article he did not mention the penis-camera specifically (ibid., 302).

The penis-camera provided researchers with a new way of seeing the interior of a woman's vagina when she was sexually aroused. At some point in the mid- to late 1950s, with some technical assistance, Masters and Johnson created the penis-camera using a see-through Plexiglas tube with a film camera and an optical eye made of plate glass inside. In another early article, Masters stated that the viewing of "the normal anatomic reactions within the vagina during the various phases of the human female's sexual response cycle have been successfully recorded in color motion-picture photography by Washington University's Department of Medical Illustration" (Masters 1960, p. 64). According to him, the penis-camera's cold-light illumination allowed the staff to observe the vaginal cavity in color clearly. Masters and Johnson described it as follows in *Human Sexual Response:*

> The artificial coital equipment was created by radio physicists. The penises are plastic and were developed with the same optics as plate glass. Cold-light illumination allows observation and recording without distortion. The equipment can be adjusted for physical variations in size, weight, and vaginal development. The rate and depth of penile thrust is initiated and controlled completely by the responding individual. As tension elevates, rapidity and depth of thrust are increased voluntarily, paralleling subjective demand. The equipment is powered electrically (Masters and Johnson 1966, p. 21).[1]

In other words, the penis-camera's "length and diameter could be selected to fit the user's vagina… The user could adjust both the depth and frequency to suit herself, and could alter the adjustments at will" (Brecher and Brecher 1966, p. 69). Thus female volunteers had total control over how the machine felt throughout the process of being filmed and over their orgasms (Masters 1960). Paul Gebhard, the director of the Kinsey Institute for Sex Research in Bloomington, Indiana, from 1956–1982, remembered that "it looked like a serious piece of medical equipment, and rather cleverly done too… You had to watch the wiring or you'd electrocute somebody. It was fairly simple, with an electric motor and a handheld rheostat to control it. It had some pretty good adjustments; otherwise it would be pretty painful" (Maier 2009, pp. 95–96).

Gebhard also had a rare opportunity to watch the penis-camera and the other machines of the sex research laboratory in action in the mid-1960s. A female subject sat in the laboratory's lounge chair, her face covered to preserve her anonymity. She was connected to an electroencephalograph machine, a heart monitor, and an electrocardiograph via wires and electrodes. Gebhard was permitted to sit close to the woman and observe the penis-camera (nicknamed "Ulysses," after the 1954 movie starring Kirk Douglas) at work, staring at the operation of the camera as the subject manipulated it between her legs. Gebhard

---

[1] The penis-camera was discarded at some point in the early 1970s after Masters and Johnson concluded their physiological research (Morrow 2008). There are no known images of it.

could watch the cervix retreat against the uterus as the woman neared climax. When the woman's movements got particularly energetic, she accidentally hit Gebhard in the eye with device, and he had to back away from her. According to Gebhard, at the end of the demonstration, Masters stated that "Males hate this machine…because invariably the females speed up the machine at a rate that no male can equal" (Maier 2009, p. 101). Lurking in Masters' comment to Gebhard was the fear that female users of the penis-camera would sideline its original scientific purpose, and that machines might replace men as providers of female sexual satisfaction. And although it was accidental, the action of the woman hitting Gebhard in the eye with the penis-camera could have a metaphoric meaning as well: that science left too little room for women to satisfy their desires on their own. For women to understand their bodies, desires, and arousal fully, men had to give them the literal and figurative space to do so.

Although Masters was vague in his early articles about the types of cameras he used and how they worked, he was clear that they provided a view of women's interior physiology that no other researcher had discovered. He also paired the film camera and penis-camera with a large electronic microscope called a colposcope to examine the activity of the clitoris during sexual stimulation (Masters and Johnson 1962, 1963a, 1964, 1966). The penis-camera in particular provided a new vision of the anatomy and physiology of women's bodies and of the new possibilities of sex research. "Colored motion-picture photography has been used to record in absolute detail all phases of the human sexual response cycle," Masters wrote (Masters 1960, p. 59). "The anatomic variations developed by the primary and secondary organs of reproduction, subsequent to effective sexual stimulation, have been successfully recorded by this most effective medium" (ibid.). Even as Masters and Johnson trumpeted their scientific measuring techniques, they recognized that "images are multiply mediated through experience, memory, and diverse modes of representation; despite their apparent close correspondence to material reality, [and] they may introduce new problems and ambiguities in the diagnostic process" (Treichler et al. 1998, p. 6). Multiple observers and measurements, they believed, would increase the accuracy of their representative findings.

Therefore, during sexual activities in the laboratory, in addition to Masters and Johnson themselves, one or two camera operators or a technician were in the room to monitor the measuring equipment. The high-quality nature of the film was due mostly to Lewis, the photographic technician. Lewis "made repeated efforts to improve the technical quality of his intravaginal photography until its fuzzy images were crystal clear" and he also made sure that "Eastman Kodak processed his films under the strictest confidentiality" (Maier 2009, p. 138). For the head researchers, technological precision equaled scientific acceptability, and the clarity of the recorded images became an indicator of the clarity of how well they were seeing scientifically. As "reading nature is a socio-cultural act," Masters and Johnson and their technicians believed that improved clarity of images would leave less room for any human error (Fausto-Sterling 2000, p. 75).

In 1960, Masters showed one of the films of women masturbating to the obstetricians and gynecologists on the Washington University medical school faculty at a Friday-afternoon seminar for the first time. The camera focused on the activity of the nipples and breasts as the woman moved through the four stages of the sexual response cycle; it never showed her face or entire body. It focused on only one part of the body at a time in order to record specific physiological changes there, such as changes to skin color. The woman's breasts were covered in electrodes, and only her torso, not her head, was visible. The doctors were startled at the high quality of the film, when they had only been accustomed (if at all) to seeing nude, sexually excited women in films that were clearly pornographic. They also came face-to-face with how little they felt they knew about female sexuality and its possible relationship to fertility (Morrow 2008; Maier 2009).

A few participants attending Masters and Johnson's professional presentations, according to a popular guide to *Human Sexual Response,* "seem to have come expecting to see a 'sex movie' or 'stag film.' They have been disappointed, of course; and some who have attended in an unscientific frame of mind have even come away disgusted with what they have seen" (Brecher and Brecher 1966, p. 81). Which aspects of films had disgusted those viewers or why is unclear—perhaps they were shocked by an up-close view of the vagina and cervix moving during sexual response and orgasm, or by the fact that women were actually having orgasms in a non-pornographic film at all. The penis-camera, which provided an interior view of women's genitalia during arousal, and the standard film camera, which showed women masturbating or men and women in coitus, both created moving images of the interiors and exteriors of sexually aroused and orgasmic women (Morrow 2008; Maier 2009). These were images that the male doctors and medical students likely had not seen on film before, had difficulty separating from their pornographic film-watching or personal experiences, or perhaps had only considered in person when they themselves were aroused during coitus with their wives or girlfriends and not giving their partners' responses their full attention. While Masters and Johnson ultimately chose to destroy their films, their tape recordings, their laboratory equipment, and most of their research notes after publishing *Human Sexual Inadequacy*—likely to preserve subject's privacy—their records of behavior ideally could have served purposes beyond their own original research (Masters and Johnson 1970).

Though Masters' original interest in the relationship of fertility and sexual behavior focused his energies primarily on women, over time he became interested in the details of men's sexuality as well. He and Johnson applied their machines to the heads, hearts, arms, and whole bodies of men—heterosexual men, that was. As he and Johnson put it in a 1963 article, "cinematography has been employed as the most effective means of recording and interpreting the finite detail of male anatomic reaction to effective sexual stimulation" (Masters and Johnson 1963b, p. 85). At the same time, they found that a minimal degree of tumescence of a woman's clitoral glans always developed under sexual tensions, though that was only viewable through a microscope. The authors observed that most male partners in a heterosexual encounter attempted "the deepest possible penetration" as their own orgasmic phase developed, rather than continuing the shallow thrusting of the plateau phase that was more likely to include labial or clitoral stimulation for their female partners (Masters and Johnson 1962, p. 256). Thus the authors were able to determine using the colposcope that the stimulation of the area around the clitoris was essential for women's stimulation to orgasm. As their male subjects paid little attention to their female partner's clitoris as they approached their own orgasmic phase, women were often left non-stimulated and non-orgasmic after a sexual encounter.

Masters and Johnson used filmic and microscopic evidence together to support their findings in order to encourage men to stimulate women near their clitorises and labias and to recognize women's needs for sexual satisfaction. By advocating such stimulation, Masters and Johnson challenged psychoanalytic, Freudian notions that women needed to transfer their orgasms from their clitorises to their vaginas to attain full maturity. Measuring different physiological processes allowed for new ways of perceiving the sexuality of bodies, as "with the very act of measuring, scientists can change the social reality they set out to quantify" (Fausto-Sterling 2000, p. 10). Machines revised and transformed knowledge about men's and women's sexual bodies in solitary and in heterosexual coition, and they produced a new level of visibility of male and female sexual bodies, interiorly and exteriorly.

What the Machines Discovered about the Sexual Body

Human Sexual Response, Masters and Johnson's book-length compilation of their physiological research, contained a rich amount of machine-derived information that challenged much received wisdom about bodies, and especially about women's sexual and reproductive bodies (Masters and Johnson 1966). It provided detailed information about women's interior pelvic anatomy that the authors had only previously published in regional medical journals directed at physicians. Among other matters, they described the four-stage sexual response cycle for men and women, confirmed the source of vaginal lubrication, demonstrated the tenting of the vagina to make room for penis and sperm, and proffered physiological data to support their contention that there were no physiological differences between vaginal and clitoral orgasms. By disproving the idea that adult women should only orgasm vaginally, they challenged contemporary psychiatric perceptions of female sexuality back to Sigmund Freud's *Three Essays on the Theory of Sexuality* and re-envisioned women's and men's bodies as spaces of sexual pleasure (Freud 1910). Ambiguities remained in their findings regarding female orgasm. While their depictions of women's sexual response cycles highlighted the importance of clitoral orgasm for a majority of women, their discovery that women with artificial vaginas could have vaginal orgasms after surgery complicated their conclusion that focusing on clitoral orgasm alone could universally solve problems of women's sexual satisfaction. Thus while the machines of Masters and Johnson's sex research led to data pointing toward the importance of clitoral orgasm, data on women after gynecological surgery showed that machine-based sex research alone could not "control, discipline, and construct the human body as a technological network of dynamic systems and forces" (Cartwright 1995, p. 3). Even in research that depended on a combination of mechanical objectivity and trained observation, ambiguity in the results still remained.

One of Masters and Johnson's best-known discoveries in *Human Sexual Response* was a four-stage response cycle, more or less matching in men and women, which created an idea that cycles of female and male sexual pleasure to orgasm could match in levels of duration and sensation (Masters and Johnson 1966). The discovery of potential or actual equality in the four-stage cycle meant that women now had clearer scientific data to support their claims to more equality in their sexual relationships. Masters and Johnson used their machine-gathered data to divide the sexual response cycle into four parts for both men and women, and tracked how the cycle affected many parts of the body. The four steps of the sexual response cycle were excitement, plateau, orgasm, and resolution.

Masters and Johnson's depiction of women's response cycle began with the changes in the excitement phase. They described the changes in each body part in the extreme detail that the film camera, penis-camera, colposcope, and their own eyes allowed them to see. In the excitement phase, their repeated observations confirmed that the nipples became erect and breasts engorged. The labia majora displaced up and away from the vaginal canal. Especially if the woman had been pregnant in the past, the labia majora and labia minora also become engorged with blood and changed color. Masters and Johnson noted that engorgement had a purpose in heterosexual intercourse: "The labia minora become so congested that they actually add a full centimeter to the functional length of the vagina by helping to provide a supportive platform for the shaft of the penis" (Masters 1960, p. 63). The "transudate-like material" that emerged from vasocongestive reactions along the vaginal wall lubricated the vagina in less than a minute after the initial physical or psychic stimulation (ibid., p. 65).

If the stimulation that produced the excitement continued into the plateau phase, the breasts became completely tumescent, and the breasts and torso often became luminescent with sweat and lightly covered with a spotty flush or rash. During the plateau phase, the inner two-thirds of the interior of a woman's vagina lengthened and distended, and the entire vagina deepened in reddish and purplish colors. The cervix elevated back and up into the lower abdomen. Heart rate increased. When the vagina was sexually unstimulated, its walls lay flat against one another; it changed in size, shape, color, viscosity, and position during sexual excitement. Mechanically mapped with the penis-camera, the vagina became a new kind of space during the arousal and plateau phases of sexual stimulation. The inner two-thirds of the vaginal barrel lengthened and distended from 2 cm wide and 7–8 cm long unstimulated to 5.75–6.25 cm wide and 9.5–10.5 cm long stimulated (ibid., p. 63). As Masters and Johnson put it, "Anatomically the vagina is a potential, rather than an actual space" (Masters 1960, p. 66) (Figs. 3.1 and 3.2).

NORMAL

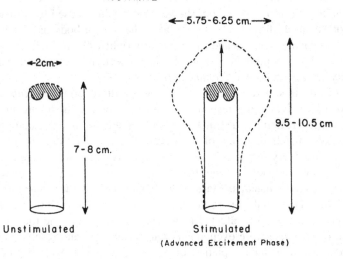

Fig. 3.1 Undistended vagina of a non-pregnant woman, stimulated and unstimulated. Masters (1960 p. 66, Fig. 4B)

SPECULUM DILATED

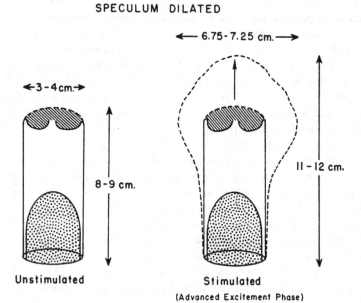

Fig. 3.2 Speculum-dilated vagina of a non-pregnant woman, stimulated and unstimulated. Masters (1960, p. 66, Fig. 4C)

Then during the brief orgasm phase, the vagina contracted strongly at roughly one-second intervals between three and ten times, and the heart rate spiked (ibid., p. 63). Masters and Johnson noted that "direct observation of the anatomic mechanism behind the orgasmic contraction response in the outer third of the vagina provides an adequate explanation for the vaginal spasm or penile grasping reaction, that has been described in general terms so many times in the literature" (ibid., p. 68). Also during orgasm, the clitoris often nearly doubled in size, the entire perineal body contracted, and the pelvis elevated spasmodically. In the resolution phase, the flush over the breasts and torso disappeared and the breasts and nipples returned to their unaroused size and shape. The vasocongestion in the labia dispersed, and the vagina slowly returned to its unstimulated state with no special coloration (Masters and Johnson 1966). Repeatedly throughout their texts, Masters and Johnson stated that their conclusions were from repeated observations with machines or the naked eye. Though the authors were confident in their findings overall, they admitted that their observations were open to error—neither machines nor the naked eye produced perfect data (Masters and Johnson 1961a, p. 107). As José van Dijck put it, "looking into a body and mapping its organic details is never an innocent act; a scan may confront people with ambiguous information, haunting dilemmas, or uncomfortable choices" (van Dijck 2005, p. 8). That statement was certainly true when Masters and Johnson evaluated their data on the sexual responsiveness of women with artificial vaginas.

Masters and Johnson affirmed the viability of artificial vaginas for female sexual pleasure—women could still have feeling and orgasms with them, ensuring that an artificial vagina was a sexual space just like "natural" vaginas were. They confirmed the "sweating phenomenon" of vaginal lubrication in a woman for whom an artificial vagina was created six years prior (Masters 1960, p. 66). They also observed lubrication in a subject who had a complete hysterectomy (removal of uterus, cervix, fallopian tubes, and ovaries) and another who had a bilateral salpingo-oopherectomy (removal of ovaries and fallopian tubes only). Masters and Johnson likewise described that women who had been ovariectomized or clitori- dectomized with full labial removal, along with postmenopausal women, were all capable of lubricating and of having orgasms if they had had orgasms before their operations. Women with artificial vaginas also had "vivid discoloration of the barrel" with the "sudden onset during the actual orgasmic experience," a "phe- nomenon of orgasmic color change" that was recorded via color cinematography (Masters and Johnson 1966, p. 108; see also Masters and Johnson 1961b). Therefore, even after major surgery on their reproductive organs and after they were no longer able to bear children, many women remained orgasmic, and their bodies remained sites of sexual pleasure.

Masters and Johnson thus argued that there was no physiological difference between vaginal and clitoral orgasm. Unlike Freudian-influenced psychiatrists, who argued that women were "frigid" if they could not transfer their ability to orgasm from their clitorises as adolescents to their vaginas as adults, Masters and Johnson provided scientific evidence that women should not feel the need to attempt that anatomically impossible feat or feel less mature or adult if they could not.

"From an anatomic point of view," Masters and Johnson wrote, "there is absolutely no difference between an orgasm stimulated by pure vaginal stimulation or pure clitoral stimulation." Regardless of the original source or type of stimulation, "the fundamental physiology of orgasmic response remains the same" (Masters and Johnson 1965, p. 528). For Masters and Johnson, psychiatrists' disavowal of the clitoris as a site for adult women's sexual pleasure denied women the possibility of considering it as a sexual space. Therefore, they confirmed with machine-gathered evidence that no parts of the exterior or interior of women's bodies should be ignored as potential sites of sexual pleasure. However, despite repeated machine and human observations, the precise combination of physical and mental sources of stimulation to orgasm, and the connections between them, remained mysterious (Fig. 3.3).

Masters and Johnson gathered data on male sexual anatomy and physiology with the same machines that they used for women, with the exception of the penis-camera. Their male material was similarly gathered from "hundreds of observations of the sexually responding human male" from film, naked-eye, and medical machinery–based sources. They pointedly mentioned that "no evaluation of the male homosexual population has been attempted during this baseline anatomic study," and reinforced their own heteronormative perspective by stating that "the heterosexual research team approach to the recording of sexual reaction has been

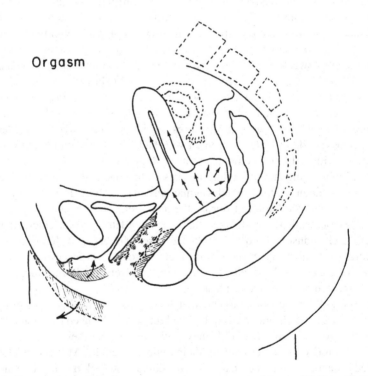

Fig. 3.3 Orgasm phase of the four-stage sexual response cycle. Masters (1960, p. 68, Fig. 6)

employed under all circumstances and at all times" (Masters and Johnson 1963b, p. 85). (They added homosexual men to their subject population later in the 1970s.) Though that was only one example of their heteronormative bias, that bias would not constrain future readings of their work as supporting sexual pleasure beyond heterosexual marriage alone.

The four-stage response cycle for men was nearly identical to that of women. Men's nipples became erect in the excitement phase, and their torsos developed a reddish flush and perspired. Penile erection in the excitement stage usually developed with extreme rapidity, usually within three to five seconds from an unstimulated, flaccid state, and the glans penis could deepen in purplish color. The scrotum tightened and the testes elevated. Men often became short of breath during the plateau and orgasmic phases, and their heart rates accelerated. The orgasmic phase involved regular recurring contractions of genital musculature and expelled seminal fluid. Masters and Johnson noted that "the frequency of the initial penile contractions has been timed at the same rate as contractions of the orgasmic platform in the vasocongested outer third of the vagina during the female's orgasmic experience" (ibid., 90). The resolution phase involved de-tumescence of the penis, and the scrotum and testes returned to their normal flaccid state.

Masters and Johnson published charts in their articles and in *Human Sexual Response* detailing the similarities of men's and women's genital and pelvic responses throughout the four stages, along with their similar changes in heart rate, breathing patterns, overall muscle contractions, and muscle tension (Masters and Johnson 1965, 1966). The one exception to men's and women's sexual similarity was that women could be re-stimulated to orgasm within seconds after a prior orgasm, but that post-orgasmic men needed to experience the resolution phase for at least a few minutes before being able to reach orgasm again. Together, the similar data and discoveries on male and female sexual responses reinforced heterosexuality. If men's and women's anatomical and physiological responses throughout the four-stage response paralleled each other, couples who were out of sync could learn how to match their responses to each other. Masters and Johnson's argument that men's and women's responses were largely equivalent, or had the potential to be, meant that women and men, as married couples, could have equal amounts of orgasms if they were able to learn how to pleasure each other properly.

When *Human Sexual Response* was published in 1966, American women's positions, both intimately and socio-culturally, were slowly beginning to change (Masters and Johnson 1966). *The Feminine Mystique,* in which Betty Friedan argued that white suburban consumer family lifestyles suffocated the women who were trying to live up to impossible feminine ideals, sparked debates across the United States about women's lives and roles in the postwar nation (Friedan 1963). The publication of *Human Sexual Response* was itself a launch pad for further public and intimate conversations about women's and men's bodies more explicitly than *Feminine Mystique:* conversations about female sexuality and orgasmic capacity, the availability and accuracy of sexual knowledge in the public sphere, and which partner in a heterosexual married couple had power and control over sexual activities and relationships. The machines that captured Masters and Johnson's data raised sociocultural questions about the place of technology in the re-envisioning of women as orgasmic and desiring sexual beings. Masters and Johnson's findings ignited public dialogues about how their technologies reformed popular cultural and psychiatric ideas about women's bodies as sexual spaces and heterosexual bedrooms as sites of non-reproductive sexual pleasure. As scientists naturally "are influenced—consciously or unconsciously—by the political needs and urgencies of their society," Masters and Johnson placed themselves in the center of intersectional debates over the role of science and technology in new understandings of women's bodies, sexuality, and pleasure (Fausto-Sterling 1992, pp. 207–208).

Criticizing the Mechanization of Sexuality

Of all the machines that Masters and Johnson used in their research, the penis-camera itself, apart from the data it gathered on the vulva, captured the imagination of many people—physicians, feminists, cultural critics, and average Americans alike. As Masters and Johnson's biographer wrote, "The totem-like quality of the mechanical dildo wasn't lost on anyone familiar with Masters and Johnson's research" (Maier 2009, p. 242). The possibility that a machine could replace a human being, and even be more satisfying sexually than one, caught people's attention as a symbol of the ways technology could change sexual and power relationships between men and women. By using the penis-camera in particular, "Masters and Johnson crafted an account of female sexuality that inadvertently threw into question the pervasive understanding of heterosexuality as innate and fully satisfied through intercourse with a penis" (ibid.). Thus Masters and Johnson's research, despite their best efforts to restrict their findings to improving the sex lives of heterosexual married couples, challenged men's and women's sexual roles. Some flesh-and-blood men were angered at the possibility that a machine could replace them sexually. In 1971's *The Prisoner of Sex,* Norman Mailer "was off on a new anger at Woman's ubiquitous plentitude of orgasms with that plastic prick, that laboratory dildoe [*sic*], that vibrator!" (Mailer 1971, p. 80). Mailer and George Ginsberg, associate director of psychiatric services at New York University Hospital, feared for the loss of male sexual privilege. Ginsberg thought that Masters and Johnson's findings about women forecast male castration anxiety and increased impotence (Allyn 2000). The 1968 film *Barbarella,* directed by Roger Vadim, showed the titular character (played by Jane Fonda) climbing into the Excessive Machine, which was supposed to stimulate her to such orgasmic heights that she would die of pleasure. Instead, the machine shreds her clothes but fails to satisfy her, overloads, and burns out. The varied reception of Masters and Johnson's machine-dependent research, from novels to movies to the general public, shows that "technology as a culture…can and must be used and consumed in a variety of ways that are not reducible to the intentions of any single source or producer" (Ross 1991, p. 131). Thus the machines of their research opened multiple possibilities for creating and envisioning new kinds of sexuality—including sexualities where men were unnecessary for women's sexual pleasure and satisfaction.

As one historian has put it, it is through Masters and Johnson's machines that "we find ourselves in the immediate presence of copulating and masturbating human beings" (Robinson 1976, p. 120). The psychiatrist Leslie H. Farber read one of Masters' first articles, "The Sexual Response Cycle of the Human Female: I," and had a particularly strong reaction to the author's use of machines, particularly the film and penis cameras, to gather sex research data (Masters 1960). He also had seen one of Masters' films and had a negative response to it. Farber respected the way that motion pictures gave "Dr. Masters the clearest edge over the subjective distortions of his predecessors" (Farber 1966, p. 297). He viewed a silent film of several women masturbating to orgasm, with only their torsos, arms, and hands visible. He watched as the anatomical and physiological changes throughout the four stages of the sexual response cycle took place across their breasts, bellies, and exterior and interior genitals. What he saw nearly revolted him: "The first shot of the moving hand heightens the dramatic effect of the oozings, engorgements, and contractions this flesh will undergo as climax approaches" (ibid., p. 298). For Farber, the female body as a color film camera captured it was not a site of pleasure or enjoyment, or even dispassionate medical or scientific interest—it was merely "this flesh," flesh that oozed, engorged, and contracted without individuality or personality. The camera turned the female body into a soulless mass of flesh, rather than a whole woman subject to a standardized medical male gaze like Farber's.

The fact that women were in control of their orgasms through the use of the penis-camera machine especially disturbed Farber. If a woman could use a machine to control the speed and timing of her orgasm, she could indulge in lustful behavior any time she wanted, and men could become unnecessary for women's sexual pleasure. Machine-made sexuality interfered with the communication of bodies and human relationships. If women turned to machines instead of men, a woman's "lust would lie to hand, ready to be invoked and consummated, in sickness or in health, in coitus or 'automanipulation,' in homosexuality or heterosexuality, in exasperation or calm, hesitancy or certainty, playfulness or despair" (Farber 1966, p. 302). For him, machine-produced sexual behavior was a scientific product, not genuine human experience, the feelings and emotions of which scientific means could never articulate.

Farber also believed that laboratory-based sex research changed the nature of bedrooms and brothels as sexual spaces, highlighting "the fundamental spatiality of all social processes"— particularly the socio-spatial dimensions of sexual behavior (Brown 2000, p. 58). Keeping those spaces apart—the first for sinful sexual experiences, the second for socially sanctioned, reproductive ones—kept the "erotic possibilities of each" high—for men, that was (Farber 1966, p. 306). When sexual behavior happened interchangeably in the laboratory, bedroom, or brothel, those spaces lost any ability to add personal meaning or resonance to the experience.

Further, when women across classes and marital statuses learned more about their bodies as sexual spaces—when they read and absorbed sexological texts—the types of experience men had in each place blurred, and, Farber seemed to imply, wives began to demand satisfactions that their husbands had left unmet. He lamented that "the emancipation which sexology enforced gradually blurred this distinction, making it unclear whether each home had become its own brothel or whether every brothel had become more like home. The truth is that sexology eventually not only blurred the distinction, but by housing us all in laboratories, made both the brothel and pornography less exciting dwellings for our erotic investigations" (Farber 1966, p. 305). Farber pointed to "the importance of sexuality in the production of space" for heterosexual male satisfaction (Brown 2000, p. 85). Cultural changes in women's sense of themselves as beings with sexual agency would change how men perceived their female partners and the spaces in which men and women had sex. If women had more control over heterosexual expressions of desire, they, rather than men, could claim homes and bedrooms as spaces where their non-reproductive sexual needs could be satisfied. For Farber, keeping married women ignorant about their bodies as sexual entities was essential to enhancing the eroticism of men's sexual experiences, and to keeping women from understanding the possibilities for their own sexual pleasures. Both antifeminist readers like Farber and feminist readers of *Human Sexual Response* had one thing in common: they believed that Masters and Johnson's work would have a significant impact on gendered and sexual relationships inside and outside the bedroom.

Masters and Johnson's research was a key inspiration for the writers of the second-wave feminism movement in the United States. Jane Gerhard has argued for the centrality of sexuality in second-wave feminist thought: "When feminists in the early 1970s sought to reimagine female sexuality, they drew on and rewrote elements of both psychoanalytic and sexologic knowledge to do so" (Gerhard 2001, p. 6). To take one example, women having orgasms with the penis-camera, as Anne Koedt argued in 1968, showed that women's sexual selves not only could survive, but could thrive without a man or male-type device. She envisioned that Masters and Johnson's "discovery" of the clitoris as the source of female orgasm, using the penis-camera and other machines in their laboratory, contained the seeds of radical political and social change. She asserted that "the recognition of clitoral orgasm as fact would threaten the heterosexual *institution*. For it would indicate that sexual pleasure was obtainable from either men *or* women, thus making heterosexuality not an absolute, but an option. It would thus open up the whole question of *human* sexual relationships beyond the confines of the present male–female role system" (Koedt 1970, emphases in original; Firestone et al. 1968). So while Masters and Johnson continually reiterated on television and in print media that their work was intended to strengthen heterosexual couples and traditional marriages, readers like Koedt noted the subversive quality of their research findings. Koedt and others "used Masters and Johnson's new research on the clitoris to overturn oppressive Freudian views of women's sexuality" (Gerhard 2001, p. 104).

Koedt was far from the only feminist to see the challenge to patriarchal social systems inherent in stating that men and women had similar capacities for and abilities to orgasm. An article in New York Radical Women's publication *Notes on the First Year*, "Women Rap about Sex," included quotations from women in two different conversations in 1968 who were frustrated with cultural tropes about female frigidity and who wanted equality, honesty, and satisfaction in their sexual relationships with men. As one participant put it, "We've got *to learn* to sleep with people because WE want them, not because THEY want us—not to prove anything to them, not in make them feel better about their masculinity, not out of weakness or inability to say no, but simply because we *want to*" (Firestone et al. 1968, emphases in original).

Some feminists made the connection between challenging patriarchy and improving women's sexual satisfaction more explicitly. Germaine Greer wrote in *Suck* magazine in 1971 that a lack of knowledge regarding the full range of women's sexual potentiality limited the findings of Masters and Johnson and the machine-captured laboratory behaviors of their research subjects. For Greer, *Human Sexual Response* was only the beginning of a genuine sexual revolution centered on women's satisfaction and pleasure: "While Masters and Johnson have done much to dispel those absurd presumptions about cunt, they could not be better than their subjects, and there is no reason why we should believe that what American middle-class women taped to electrodes could do, is all that could have been done" (Heidnry 1997, pp. 141–142; Masters and Johnson 1966; Gerhard 2001). In the vision of women's liberation put forth by Koedt, Greer, Ti-Grace Atkinson, Kate Millet, Shulamith Firestone, and others, if more women fully understood their bodies' capacities for pleasure, they might not need men or marriage at all, thus upturning centuries of social organization around heterosexual marriage and family that ignored, or at least downplayed, women's sexual needs. Together, second-wave "feminists offered a counterhegemonic model of female sexuality by infusing it with the new values of self-determination, autonomy, and equality" (Gerhard 2001, p. 115; see also Millet 1970; Greer 1970; Firestone 1970; Faludi 2013).

Virginia Johnson tried to combat the notion that her work with Masters had potential feminist qualities. She argued that "lady liberationists" and "emancipated women" were a danger to society because they made men feel sexually inadequate (Allyn 2000, p. 169). Johnson felt that heterosexual women who manifested feminist ideas in the bedroom could cause sexual dysfunction in men, and she did not want her and Masters' work read as a document for women's sexual liberation. Nonetheless, "that scientists might be unaware of the implications of their work does not make them any less mediators or marketers of political ideas," and so despite their public efforts, Masters and Johnson could not control how their readers thought about and used their work (Schiebinger 2004, p. 8). And their readers, especially their feminist and homosexual ones, saw the liberating potential of their work and began to incorporate it into their own thinking and activism.

Conclusion

As Masters and Johnson unequivocally named the vaginal orgasm and the clitoral orgasm as the same entity (even if they were not sure how arousal happened in the first place), they overturned decades of psychological and psychiatric scholarship affirming that women needed to move the site of their juvenile orgasms, the clitoris, to the site where adult women had them, in the vaginal canal. Women had heard from Freud onwards from medical professionals that if they did not manifest "orgasm transfer," they were "frigid," and needed psychiatric care if they were not able to effect such a transfer on their own. Masters and Johnson proved with their machines that the clitoris was the place on the woman's body most richly supplied with nerves, and that telling women they needed to transfer the site of their orgasm was medically and scientifically unsupportable.

There were, of course, negative aspects to Masters and Johnson's findings. They presented and enforced married heterosexuality as normal; defined homosexuality as abnormal; promoted "biological reductionism" and "false biological uniformity"; and they only used volunteers who had successfully and consistently achieved orgasm, and who were able to perform sexually in highly unusual conditions (Tiefer 1995, p. 57). Some scholars have seen their articulation of a matching four-stage response cycle for men and women as artificial, forcing couples into expecting sexual symmetry despite biological and sociocultural differences between the sexes and fostering disappointment when that did not happen (Tiefer 1995; Morrow 2008). Nevertheless, "the kinds of liberatory fantasies that surround new technologies are a powerful and persuasive means of social agency," and the technologies that Masters and Johnson used forwarded the sexually emancipatory aims of second-wave feminism (Penley and Ross 1991, p. xii).

Despite the many problems with Masters and Johnson's research, participants in the sexual revolution of the late 1960s and beyond found their research liberating underneath the heterosexist language. For example, a lesbian graduate student in physical therapy wrote to Johnson in 1975: "As a feminist I must thank you for the extensive involvement of women in your research. Your work has freed many women that the 'Sexual Revolution' neglected" (Petrowski 1975). Another woman wrote to Johnson in 1970, chiding her for stance that women should express themselves sexually only in marriage and asking her to not "perpetrate in any way the double standard which has been such an obstacle to satisfying sexual relationships for so long, but instead [to] support the right of every woman to attain full sexuality, by whatever method is best for her" (Chernow 1970). Masters and Johnson's research had implications beyond, and often contrary to, their own expectations for reconfigurations of women's sense of sexual identity and agency.

Masters and Johnson's understanding of male and female physiology likewise had multiple spatial effects beyond their original research agenda: they identified and prioritized the female vulva's sexual qualities and potentialities as a space of pleasure above its culturally preferred reproductive qualities, and they positioned female orgasm as the physiological equivalent of male orgasm. Their research also led to new considerations of private homes as spaces for men's and women's sexual pleasure apart from spaces where sexual reproduction and male satisfaction alone took place. Masters and Johnson's machines as a group and their users provided men and women measured evidence of the physiological aspects of their orgasms, even if neither the machines nor their operators could solve the question of what stimulated women's arousal. They gave heterosexual women new scientific grounding for claiming orgasmic equality in their sexual relationships with men. Those new spaces—inside women and inside their relationships—became visible with the interlinked results of the multiple machines that Masters and Johnson used to record their data.

Finally, *Human Sexual Response* became an inspirational text for a new generation of sex researchers who were interested in continuing Masters and Johnson's initial investigations into sexual physiology (Masters and Johnson 1966). Masters and Johnson's well-publicized use of new, creative machines such as the penis-camera, and standard medical equipment such as the respirator, in their research made technology central to sex research. The idea that researchers could measure aspects of women's arousal from the interior of the body would galvanize a small but significant portion of the nascent sexological research community in the late 1960s and 1970s. After the mid-1970s, when aversion therapy for machines had been discredited and Masters and Johnson had moved on to marriage counseling, scientific researchers began to use and to develop different machines to compare and to contrast more seriously the physiological aspects of men's and women's sexual arousal and response.

References

Allyn, D. 2000. *Make love, not war: The sexual revolution; an unfettered history.* Boston: Little, Brown.

Brecher, R., and E. Brecher (eds.). 1966. *An analysis of Human Sexual Response.* Boston: Little, Brown.

Brown, M.P. 2000. *Closet space: Geographies of metaphor from the body to the globe.* London: Routledge.

Bullough, V.L. 1994. *Science in the bedroom: A history of sex research.* New York: Basic Books.

Cartwright, L. 1995. *Screening the body: Tracing medicine's visual culture.* Minneapolis: University of Minnesota Press.

Chernow, C., and V.E. Johnson. 1970. *"To Be Filed" folder 2, box 5, Virginia Johnson Masters Collection.* Bloomington, Indiana, USA: Kinsey Institute for Research in Sex, Gender, and Reproduction Inc.

Davidson, J.O., and D. Layder. 1994. *Methods, sex, and madness.* London: Routledge.

D'Emilio, J., and E.B. Freedman. 1988. *Intimate matters: A history of sexuality in America.* New York: Harper & Row.

van Dijck, J. 2005. *The transparent body: A cultural analysis of medical imaging.* Seattle: University of Washington Press.

Ericksen, J.A., and S.A. Steffen. 1999. *Kiss and tell: Surveying sex in the twentieth century.* Cambridge: Harvard University Press.

Faludi, S. 2013. Death of a revolutionary. New Yorker (15 Apr 2013). http://www.newyorker.com/reporting/2013/04/15/130415fa_fact_faludi. Accessed 13 Apr 2013.

Farber, L.H. 1966. I'm sorry, dear. In *An analysis of human sexual response*, ed. R. Brecher, and E. Brecher, 291–318. Boston: Little, Brown.

Fausto-Sterling, A. 1992. *Myths of gender: Biological theories about women and men*, 2nd ed. New York: Basic Books.

Fausto-Sterling, A. 2000. *Sexing the body: Gender politics and the construction of sexuality*. New York: Basic Books.

Firestone, S. 1970. *The dialectic of sex: The case for feminist revolution*. New York: Morrow.

Firestone, S., et al. 1968. Notes from the first year. New York Radical Women, New York. http://library.duke.edu/rubenstein/scriptorium/wlm/notes. Accessed 13 Apr 2013.

Friedan, B. 1963. *The feminine mystique*. New York: Norton.

Freud, S. 1910. *Drei abhandlungen zur sexual theorie [Three essays on the theory of sexuality]*. Leipzig: Franz Deuticke.

Gerhard, J. 2001. *Desiring revolution: Second-wave feminism and the rewriting of American sexual thought, 1920 to 1982*. New York: Columbia University Press.

Greer, G. 1970. *The female eunuch*. London: MacGibbon & Kee.

Heidenry, J. 1997. *What wild ecstasy: The rise and fall of the sexual revolution*. New York: Simon & Schuster.

Irvine, J.M. 2005. *Disorders of desire: Sexuality and gender in modern American sexology*. Rev ed. Philadelphia: Temple University Press.

Kinsey, A.C., W.B. Pomeroy, and C.E. Martin. 1948. *Sexual behavior in the human male*. Philadelphia: W. B. Saunders.

Kinsey, A.C., et al. 1953. *Sexual behavior in the human female*. Philadelphia: W. B. Saunders.

Koedt, A. 1970. The myth of the vaginal orgasm. http://www.uic.edu/orgs/cwluherstory/CWLUArchive/vaginalmyth.html. Accessed 22 Mar 2013.

Lefebvre, H. 1991. *The production of space*. Trans: D. Nicholson-Smith. Oxford: Blackwell.

Maier, T. 2009. *Masters of sex: William Masters and Virginia Johnson, the couple who taught America how to love*. New York: Basic Books.

Mailer, N. 1971. *The prisoner of sex*. Boston: Little, Brown.

Maines, R.P. 1999. *The technology of orgasm: "Hysteria," the vibrator, and women's sexual satisfaction*. Baltimore: Johns Hopkins University Press.

Masters, W.H. 1959. The sexual response cycle of the human female: II. Vaginal lubrication. *Annals of the New York Academy of Sciences* 83: 301–317. doi:10.1111/j.1749-6632.1960.tb40904.x.

Masters, W.H. 1960. The sexual response cycle of the human female: I. Gross anatomic considerations. *Western Journal of Surgery, Obstetrics, and Gynecology* 68(1): 57–72.

Masters, W.H., and V.E. Johnson. 1961a. The physiology of the vaginal reproductive function. *Western Journal of Surgery, Obstetrics, and Gynecology* 69(2): 105–120.

Masters, W.H., and V.E. Johnson. 1961b. The artificial vagina: Anatomic, physiologic, psychosexual function. *Western Journal of Surgery, Obstetrics, and Gynecology* 69(3): 192–212.

Masters, W.H., and V.E. Johnson. 1962. The sexual response cycle of the human female: III. The clitoris: Anatomic and clinical considerations. *Western Journal of Surgery, Obstetrics, and Gynecology* 70(5): 248–257.

Masters, W.H., and V.E. Johnson. 1963a. Clitoris: An anatomic baseline for behavioral investigation. In *Determinants of human sexual behavior. Charles C. Thomas*, ed. G. Winokur, 44–51. Illinois: Springfield.

Masters, W.H., and V.E. Johnson. 1963b. Sexual response of the human male: I. Gross anatomic considerations. *Western Journal of Surgery, Obstetrics, and Gynecology*, 71(2): 85–95.

Masters, W.H., and V.E. Johnson. 1964. Sexual response: Part II, anatomy and physiology. In *Human reproduction and sexual behavior*, ed. C.W. Lloyd, 462–473. Philadelphia: Lea and Febiger.

Masters, W.H., and V.E. Johnson. 1965. Sexual response cycles of the human male and female: Comparative anatomy and physiology. In *Sex and behavior*, ed. F.A. Beach, 512–534. New York: Wiley.

Masters, W.H., and V.E. Johnson. 1966. *Human sexual response*. Boston: Little, Brown.

Masters, W.H., and V.E. Johnson. 1970. *Human sexual inadequacy*. Boston: Little, Brown.

Millet, K. 1970. *Sexual politics*. New York: Ballantine.

Morrow, R. 2008. *Sex research and sex therapy: A sociological analysis of Masters and Johnson*. New York: Routledge.

Penley, C., and A. Ross (eds.). 1991. *Technoculture*. Minneapolis: University of Minnesota Press.

Petrowski, M., and V.E. Johnson. 1975. *"To Be Filed" folder 2, box 5, Virginia Johnson Masters Collection*. Indiana, USA: Kinsey Institute for Research in Sex, Gender, and Reproduction Inc., Bloomington.

Reumann, M.G. 2005. *American sexual character: Sex, gender, and national identity in the Kinsey reports*. Berkeley: University of California Press.

Robinson, P. 1976. *The modernization of sex: Havelock Ellis, Alfred Kinsey, William Masters, and Virginia Johnson*. New York: Harper & Row.

Ross, A. 1991. Hacking away at the counterculture. In *Technoculture*, ed. C. Penley, and A. Ross, 107–134. Minneapolis: University of Minnesota Press.

Schiebinger, L. 2004. *Nature's body: Gender in the making of modern science*, New ed. New Brunswick: Rutgers University Press.

Tiefer, L. 1995. *Sex is not a natural act and other essays*. Boulder, Colorado: Westview Press.

Treichler, P.A., Cartwright L., and Penley, C. 1998. Introduction. In *The visible woman: Imaging technologies, gender, and science*, ed. P.A. Treichler, L. Cartwright, and C. Penley, 1–18. New York: New York University Press.

Chapter 4
The Vaginal Photoplethysmograph and Devices for Women: Gauging Female Arousal

Abstract Researchers began to develop and to design machines for measuring women's physiological changes during sexual arousal in the context of second-wave feminism in the late 1960s and early 1970s. A number of women-oriented machines were only used once or twice due to their painfulness, impracticality, or inability to provide quality data for analysis. Machines for women that have achieved stability in the research community include the vaginal photoplethysmograph (aka vagina photometer) and the labial thermistor. The vaginal photoplethysmograph measures vaginal blood volume (VBV) or vaginal pulse amplitude (VPA), depending on whether the researcher uses DC or AC current, respectively. Though the vaginal photoplethysmograph is currently the most popular instrument in studies of women's sexualities, the fact that no one is sure what it measures or what those measurements indicate hampers its usefulness. In light of the problems associated with the vaginal photoplethysmograph, researchers often put it or other genital measurements together with non-genitally specific measurements such as blood pressure cuffs, respirators, and thermographs. Those combinations of machine-gathered data over the past four decades show that overall men's subjective and physiological patterns of sexual arousal correlate much more often than do women's patterns. Whether those gender differences are physiologically based or due to results from incomparable machines remains an open question.

Keywords Vaginal photoplethysmograph · Vaginal photometer · Labial thermistor · Sex research · Sexual arousal · Sexuality

As academic interest in sex research grew in the postwar scientific era, so too did the professional organizations and journals that supported sexology as a legitimate field. Meetings for the academic Society for the Scientific Study of Sex (SSSS) began in 1957 in New York City, and the organization was formalized in 1960. The Sexuality Information and Education Council of the United States (SIECUS), organized to provide sex information to the wider American public, was founded in 1964 (Bullough 1994). Richard Green was the founder and first editor of the *Archives of Sexual Behavior* in 1971, which became the journal of record for

D. J. Drucker, *The Machines of Sex Research*,
SpringerBriefs in History of Science and Technology,
DOI: 10.1007/978-94-007-7064-5_4, © The Author(s) 2014

sexology, and he led the organization of the first meeting of the International Academy of Sex Research in Stony Brook, New York in 1975 (Green 1971, 1985; Bullough 1994). The creation and use of specific machines to measure aspects of female sexuality, including the vaginal photoplethysmograph, coincides with the development of sex research as an academic discipline. From the time of the publication of *Human Sexual Response* in 1966 onward, professional academic interest in human sexual physiology and women's sexual physiology in particular began to grow (Masters and Johnson 1966). For academics curious about exploring William Masters and Virginia Johnson's findings in more depth, the late 1960s and 1970s were a period of experiment and exploration.

The availability of a birth control pill in the United States beginning in 1960 also contextualized change in scientific thinking on the part of researchers interested in sexual physiology (Tone 2001; Gordon 2002; Watkins 2001). As Elaine Tyler May writes, "the pill made it possible, for the first time ever, to separate contraception from the act of intercourse" (May 2010, p. 57). The pill made contraception invisible during sexual intercourse, unlike the condom, diaphragm, or using withdrawal as a contraceptive method. The pill made it easier for scientists to see and to study women's sexual arousal, desire, and behavior as separate from their reproductive capacities. Although "the availability of a particular technology rarely creates an immediate change in deeply held cultural values," May argues, "the pill was a part of a changing cultural environment that was gradually becoming more tolerant" (ibid., p. 73). That more tolerant cultural environment influenced medical and scientific communities as well. Professional, technological, and cultures together made the possibility of basic research on women's sexual anatomy and physiology into a reality.

The professionalization of human sex research on women took place in a rapidly shifting academic, legal, and cultural landscape. Hormonal birth control was first approved by the U.S. Food and Drug Administration (FDA) in 1960, it reached West Germany in 1961 and East Germany in 1965, and millions of women around the world used it by the end of the decade (May 2010). Women's legal rights regarding their sexual and reproductive rights were undergoing radical changes. In 1965, the U.S. Supreme Court upheld the right of married people to use artificial birth control in *Griswold v. Connecticut,* and unmarried people gained the legal rights to possess birth control in *Eisenstadt v. Baird* (1972) (Johnson 2005; Mims 1972). A woman's right to have an abortion was legalized in January 1973 following the *Roe v. Wade* Supreme Court decision. However, as of 1974, the American second-wave feminist movement as a whole was experiencing difficulties, due in part to the defeat of the nationwide campaign for an Equal Rights Amendment to the U.S. Constitution (Mansbridge 1986; Berry 1988). Additionally, the movement was beginning to fragment over issues such as race, class, sexual identity, and lesbian and radical separatism (Gerhard 2001). Nevertheless, second-wave feminism had already made a lasting impression on academic sex research. A central insight of radical feminists was that "the value of women's sexual self-determination [served] as the foundation for feminist politics" (Gerhard 2001, p. 193). An improved ability of women to manifest their sexual self-determination became one motivation for sex researchers in both academic and clinical settings.

Moreover, researchers were aware of the cultural changes surrounding women's rights and women's sexualities, and felt the impact of those changes in their everyday professional lives. Wrote one in a review of literature on women's sexuality, "An awareness of women's political, personal, and social rights has been a relatively recent development in western culture, and this awareness is currently a strong impetus for sex researchers and clinicians" (Hoon 1984, p. 768). Researchers with an interest in basic sexual physiology not only received inspiration from the global women's movement, they were also willing to devote time, energy, and resources into science that could manifest into new and better understandings of women's bodies as sexual entities beyond their reproductive capacities. Interest in research on women's sexual physiology reached a peak of academic interest between 1974 and 1984. From the mid-1980s onward, the field of sex research as a whole broadened outward into multiple subfields.

By the early 1980s, scattered sex researchers, including those with a special focus on physiology, were attempting to articulate a broader theoretical background for their discipline. In 1981, two psychologists at the University of California, Los Angeles (UCLA) reviewed contemporary methodological considerations of sex research. They conceived of sex research still as a "field" that "spans numerous scientific disciplines" rather than a discipline of its own (Bentler and Abramson 1981, p. 226). They pointed out that workers in the field adopted the practices—and not always the best practices—of their home discipline, and they did not always concur on what the best practices were (ibid.). Some subfields of sex research were so new—such as research centered on rape—that no one had yet developed research protocols, let alone agreed on the specific research problems, hypotheses and theories. So sex researchers focused on women, and on developing machines for research on women, were creating tools and the groundwork for a new field and a new discipline. As Bentler and Abramson wrote, "techniques and methodologies may be used to gain an initial handle on a phenomenon that has not been well measured, systematically observed, or placed into broader research contexts" (ibid., p. 227). The creation and use of specific instruments and developing consensus around them were methods of providing unity in a field that was, at times, frustratingly interdisciplinary and fragmented.

Measure for Measure: Inventing Machines for Female Sexual Response

Beginning in the late 1960s, and following the academic impact of the widely read *Human Sexual Response,* researchers concerned with female sexuality experimented with a wide variety of devices to investigate women's sexual physiology (Masters and Johnson 1966). Julia R. Heiman, now director of the Kinsey Institute for Sex, Gender, and Reproduction in Bloomington, Indiana, specifically remembered reading *Human Sexual Response* and thinking of it as a starting point for exploring gender similarities and differences in human sexual behavior (Masters and Johnson 1966; Heiman 2013). She also recalled the influence of Helen Singer Kaplan on scholarly interest in women's sexuality, though Kaplan was a psychoanalyst and did not use machines in her work (Bergner 2009; Kaplan 1974). Although no other researchers followed Masters and Johnson's lead in further refining the technology of the penis-camera, the idea of focusing scholarly attention on women, and on using technology to map the interiors of women's bodies scientifically, had also taken hold in the imaginations of other scientists.

Some early machines and processes were so painful for voluntary female subjects that out of charity for the subject—or disappointment with the results—the machines would immediately disappear from the landscape of medical journals. Two sets of researchers, one in the U.S. and one in Germany, used imaginative methods in order to consider the role of uterine contractions in sexual arousal. They attached a rubber balloon to a polyethylene tube, which was placed inside the uterus (probably using a speculum) and filled with water. The researchers then measured the pressure of contractions on the balloon with a transducer attached to a recorder, but female subjects regularly expelled the balloon on account of extreme pain (Bardwick and Behrman 1967; Jovanovic 1971). As a reviewer dryly noted, the "insertion of a balloon in the cervix is quite painful for many women which would limit the usefulness of this technique" (Zuckerman 1971, p. 316). It is hard to imagine the discomfort that some subjects must have felt when trying to stay aroused during such procedures (Fig. 4.1).

Researchers also explored the idea of measuring women's sexual arousal via changes in bodily or specifically genital temperatures. Seymour Fisher and Howard Osofsky, physicians at the Upstate Medical Center in Syracuse, New York, developed a device that took a woman's temperature inside her vagina (Fisher and Osofsky 1967, 1968). Masters and Johnson had used electrodes to

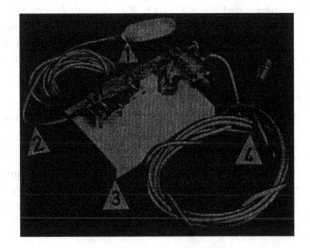

Fig. 4.1 Photograph of supplementary apparatus for the recording of vaginal contractions. (*1*) Rubber balloon, (*2*) plastic tube between rubber balloon and pressure chamber, (*3*) pressure chamber with water container, and (*4*) plastic tube from pressure chamber to EEG channel. Jovanovic (1971), p. 318, Fig. 10. Reprinted with kind permission from Springer Science + Business Media, copyright 1972

measure the pH, but not the temperature, of female subjects' vaginas earlier in the 1960s (Masters and Johnson 1961a, b). Fisher and Osofsky drew their inspiration from *Sexual Behavior in the Human Female* and from *Human Sexual Response*, as those were the only two citations in their 1968 article aside from their own work (Kinsey et al. 1953; Masters and Johnson 1966; Fisher and Osofsky 1968). They were interested in the relative temperatures in different parts of a woman's body during sexual arousal, and the ways that those temperatures correlated, or did not correlate, with subjective measurements of women's emotions about their orgasms in interviews.

The forty-two married women in their study, in three separate recordings, had temperature measurements attached to their ankles, fingers, labia, and breasts, and Yellow Springs Thermistor Probes inserted in their rectums and vaginas for a total of twenty minutes (Fisher and Osofsky 1968). The researchers also measured subjects' heart rate and skin resistance. The second and third measurement sessions included some discussion of the subjects' experiences of intercourse and orgasm, a partial gynecological exam, and some time alone, during which the subject could presumably masturbate. Unfortunately, the description of the measurement sessions is unclear on that point, although it seems that subjects were offered the opportunity for self-stimulation, and the researchers tracked their measurements while they did so (ibid.). The researchers concluded, "The higher the vaginal temperature the greater is the tendency to experience orgasm as a highly aroused, unusual state and to be unaware of feelings of weakness or tiredness" (ibid., p. 223). The rectal and vaginal temperatures were similar, as were the ankle and finger temperatures, so all of those temperature measurements provided little new information about women's sexual bodies. Thus in light of the relatively unremarkable results using standardized temperature probes in the vagina and rectum—they were simply a more invasive means of measuring core body temperature—other sex researchers did not pursue genital temperature measurements for another decade. Obstetricians and gynecologists instead focused their attention on the measurement of vaginal blood flow.

In 1965, two researchers in Israel, Yoram Palti and Bruno Bercovici, used new light-reflective technologies to create the first vaginal photoplethysmograph (aka vaginal photometer) (Palti and Bercovici 1967; Bercovici and Palti 1967). Bercovici was an obstetrician and gynecologist at the Hebrew University-Hadassah Medical School. Palti and Bercovici's aim in their research project was to measure blood flow in women's genitals to better understand their vascular systems, as researchers had previously done work in that area only on nonhuman female animals. Their vaginal photoplethysmograph, attached to the end of a standard speculum, captured the amount of reflected light in vaginal tissues using a photoelectric cell assembly with AC current. A miniature electric light shone on the upper vaginal walls, and the light then reflected back to the photoelectric cell. The variations in reflected light due to the subject's pulse rate were converted into "electric potential variations" (a triangulated measurement of pulse height in millimeters) and then recorded on an electrocardiograph (Palti and Bercovici 1967, p. 144). In a second experiment, they modified their device to capture pulse waves in both the upper and lower vagina, and compared them both with pulse waves of the finger (Bercovici and Palti 1967). They called the measurement "vaginal pulse amplitude" (VPA) (Figs. 4.2, 4.3).

Palti and Bercovici were interested in finding changes in pulse amplitude across the menstrual cycle to understand better the interplay of women's vascular and hormonal changes throughout it. They found for women without gynecological disorders, that pulse amplitude was highest around ovulation and around the twenty-first day of a woman's menstrual cycle, though the reasons were unclear (Palti and Bercovici 1967). They did not suggest in either article that other physicians could adopt their device for other purposes, such as measuring vaginal blood flow to study sexual physiology. After working together on the vaginal photoplethysmograph, Bercovici and Palti collaborated once more in 1972, investigating the effects of estrogens on cytology, but after that they no longer published together (Bercovici et al. 1972). Bercovici moved into studying vaginal cytology after developing the device and did not pursue it further (Bercovici et al. 1973). Palti continued to study blood vessels using plethysmography in animals and other parts of the human body, but not in the vagina (Uretzky and Palti 1977). As their vaginal photoplethysmograph was only used for two small experiments in Jerusalem, and they presented their work at only one conference, most physiologists did not know about the device or were uninterested in developing it themselves.

Fig. 4.2 A vaginal speculum used for recording the pulse waves from the upper and lower vagina. Bercovici and Palti (1967), p. 415, Fig. 1. Copyright Elsevier, 1967

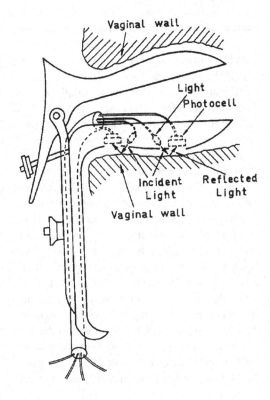

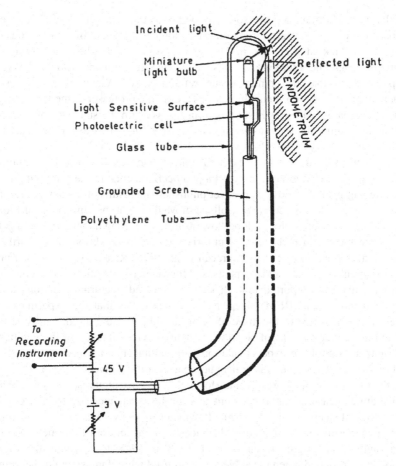

Fig. 4.3 A pulse detector for recording the endometrial pulse. Bercovici and Palti (1967), p. 416, Fig. 3. Copyright Elsevier 1967

American researchers also found inspiration from *Human Sexual Response* for new machines that could measure aspects of women's sexual physiology (Masters and Johnson 1966). Many, if not most of those machines, were one-time experiments that were not even the subject of a full academic article, just descriptions of conference presentations or records of personal conversations of experimenters with a journal author. In most cases, no photographs or drawings of them exist, so readers must use their imaginations to recreate what such devices looked like. The diversity of machines described in those brief accounts shows how creative sex researchers and their technicians were in thinking about how to capture data regarding women's sexual physiology, not to mention how patient their subjects were with being fit with strange assemblies. That diversity also illustrates that researchers were in many respects still unsure of what they were actually measuring, and of what they hoped to find about women's sexual desire, arousal, or response.

Marvin Zuckerman, a professor of psychology at the University of Delaware, in preparing a comprehensive article on physiological measures of sexual arousal in 1969–1970, tackled that task in two ways. Not only did he read through all available academic literature, he also contacted researchers who had work in preparation so that he could include references to their work in his article (Zuckerman 1971). He found Charles T. Tart, then a professor at the University of California, Davis, who was best known for his work on the state of consciousness (particularly drug-altered consciousness). In early 1969, Tart experimented with a "clitorophotoplethysmograph," which measured blood flow in the clitoris (ibid., p. 315). Part of the device surrounded the clitoris and was mounted on a stabilizing rod that fit into the vagina. It also had a photocell attached to the stabilizing rod to measure vaginal blood flow. Tart never published the results of his device. Another group of researchers based at the College of Medicine at the University of Florida attempted to measure clitoral blood flow during sleep using a mercury strain gauge on two women who had enlarged clitorises due to a condition called virilizing congenital adrenal hyperplasia (Karacan et al. 1970; Roach 2008). In measuring those women's clitorises and the penises of four men (two with the same condition, two without), they reported that all subjects showed nocturnal erections during rapid-eye movement (REM) sleep. No evidence exists that the experiment was repeated, perhaps due to the discomfort of the subjects, and the difficulty of their achieving regular patterns of sleep with such a device attached to the genitals.

Another example of experimentation with genital measurement machines was that of Arthur Shapiro and his colleagues, based at the University of Pennsylvania and the Institute of the Pennsylvania Hospital, who presented research on a vaginal temperature measurement instrument at a professional meeting for the psycho-physiology of sleep. It was "a thermal flowmeter for the vaginal wall...constructed using two thermistors held at equal distances from the anterior mid line by the ring of a vaginal diaphragm" (Shapiro et al. 1968, p. 394). They found that vaginal temperature dropped when the subject was aroused while fantasizing or hearing an audiotape of erotic literature. They focused on their measurement of one subject's arousal during sleep. Except for the University of Florida experiment described above, and the earliest recorded use of the penile strain gauge in Germany, this experiment was only third published example of sex research that involved using machines on sleeping subjects (Ohlmeyer and Brilmayer 1947; Ohlmeyer et al. 1944). The subject of this experiment had seven REM periods overnight, and three of those had at least one episode of increased vaginal blood flow. Seven episodes of increased vaginal blood flow occurred in part during non-rapid eye movement (NREM) sleep, showing that at least one female subject recorded similar arousal patterns to Ohlmeyer's subjects (Ohlmeyer et al. 1944).

Shapiro's colleague Harvey D. Cohen presented a follow-up study on four women's sexual arousal patterns in a waking state at a psychophysiological research meeting in 1970. Vaginal blood flow increased when women fantasized sexually, but not when they imagined being in a state of fear or anxiety (Cohen and Shapiro 1971). The diaphragm-based flowmeter was never developed further, in part due to Shapiro's death, although Cohen remained involved in sex research through the mid-1970s (Cohen et al. 1976). A further complication was that each device had to be fashioned and fitted for each subject, adding markedly to the time and expense of any experiment using it. The results from the thermal flowmeter were only presented at conferences and were never published in an academic journal, so they received limited academic attention and did not catch on as a device for others to build on and to use.

The Vaginal Photoplethysmograph, the Labial Clip, and the Thermograph

It took seven years after the Bercovici-Palti studies were published for others to invent a more subject-friendly vaginal photoplethysmograph for studies that explored the relationship of vaginal physiology to sexual arousal and desire. Although the penis-camera was central to William Masters and Virginia Johnson's research on women's sexual physiology, Masters and Johnson did not keep the device after their turn to marriage therapy (Masters and Johnson 1966). No photographs or detailed descriptions of it existed either, making it difficult for anyone else to recreate it (Morrow 2008; Maier 2009). The next version of a vaginal measurement device, the vaginal photoplethysmograph, was created in the department of psychology at the State University of New York (SUNY), Stony Brook, in 1974. The device was invented in the psychophysiology laboratory of James H. Geer, who in 1971 was working on projects with graduate students on fear and responses to fear. While one of his male graduate students easily set up a polygraph to measure penile responding, Geer remembered, "women working in the laboratory complained that our studies were severely restricted by being unable to study women and lobbied for looking at methods to study responding in women" (Geer 2005, p. 285). As Geer read literature on photoplethysmography and thought about how he could apply it to women, he could not figure out a suitable placement for the light source. After he mentioned the problem to his wife, she suggested that the device be tampon-shaped, as women would be familiar with the sensation of having a device of that size and shape in their vaginas. Women subjects would also know how to insert and to remove it themselves. Julia R. Heiman, then a graduate student in Geer's laboratory, remembered that female graduate students in the laboratory also advised Geer on the form of the device (Heiman 2013).

Geer, with the assistance of George Sintchak, a laboratory technician in the SUNY Stony Brook psychology department, created such a tampon-shaped device (4.5 mm long and 1.2 mm in diameter) made of clear acrylic plastic (Sintchak and Geer 1975). They were only retroactively aware of the Palti-Bercovici design, which required that a physician or nurse hold the speculum in place during measurement (Geer 2005). Heiman recalled that the earliest designs tended to break, and that the graduate students and laboratory technicians also had to figure out protocols for cleansing and disinfection, since multiple subjects would use each device (Heiman 2013). Geer's final iteration of the vaginal photoplethysmograph, which is still the most popular, contains a very small lamp and photocell, which were then connected to a plug box via a bridge circuit and shielded cables. It measures the reflected light inside the vaginal walls in vaginal blood volume (VBV) with DC current or vaginal pulse amplitude (VPA) with AC current (Geer et al. 1974; Geer 1975; Janssen et al. 2007). Vaginal blood volume showed an estimate of the total amount of blood in the vagina, usually measured once per second for three to five intervals of twelve seconds each (Fig. 4.4).

Geer and his junior colleagues tested the new instrument with twenty female undergraduate students at their university. After the subjects had an initial meeting with two female researchers for an introduction to the study, they returned to the laboratory a second time for the actual study. The researchers asked them to cleanse the vaginal photometer and showed them to a private room, where the subjects inserted the device into their vaginas and watched a non-erotic test film along with a stag film that the researchers had used previously with male subjects (Geer et al. 1974). They thought initially that women would not have any response to the stag films, but the women had a near-instant response to it, so they immediately designed and carried out the first study demonstrating how well the device worked to measure women's physiological arousal (Geer 2005). As Geer and Sintchak stated in their article describing the mechanics of the vaginal photoplethysmograph, "research on sexual arousal in women has been restricted by the lack of a simple, reliable, yet inoffensive procedure for assessing vaginal blood volume. The described device provides a partial solution to that problem" (Sintchak and Geer, 1975, p. 115).

The vaginal photometer was able to measure aspects of a woman's sexual physiology that William Masters and Virginia Johnson's less-sensitive penis

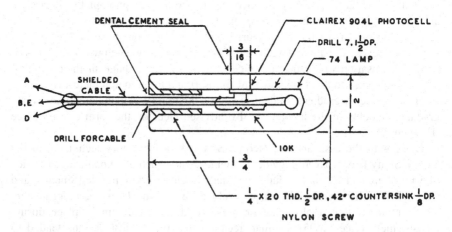

Fig. 4.4 Detail drawing of a vaginal plethysmograph probe. Sintchak and Geer (1975), p. 114, Fig. 1. Reprinted with permission from John Wiley & Sons, copyright 2007

camera could not (Masters and Johnson 1966). In experiments, the genital devices were used along with other kinds of machines. Experimenters alternated neutral, violent, and erotic audiotapes and videotapes to investigate whether VPA and/or VBV had some relationship to sexual arousal, a similar process as previous researchers had used with other instruments. Such interviews were combined with personal interviews and written questionnaires to compare physiological measurements with subjects' feelings about and perceptions of their arousal. In early experiments, researchers found that VPA related more directly to sexual arousal than did VBV (Heiman 1977; Heiman and Maravilla 2007). Scientists in the present generally consider VPA a more reliable measure of arousal, though some studies have shown that VBV and VPA are only sometimes connected to subjective feelings of arousal (Janssen et al. 2007; Woodward and Diamond 2009). Nonetheless, some kind of genital instrument is now essential to sex research, and it is usually combined with one or more non-genital instruments.

A replication of the American vaginal photoplethysmograph experiments by a Canadian research team using six female subjects confirmed that the device measured vaginal blood flow related to sexual arousal. Furthermore, it had the most sensitivity to physiological changes among the six devices they used on subjects all at once (including forehead temperature taken with a thermometer, a blood pressure cuff, electrodes on the feet for skin sensitivity, and blood pulse amplitude and heart rate inside a transducer on the finger) (Hoon et al. 1976). Unlike the blood pressure cuff and thermometer in particular, the vaginal photoplethysmograph was not uncomfortable or awkward, and "several women have needed to be reminded that the probe was in their vaginas at the end of clinical or experimental sessions, so comfort is clearly not a problem with this measure" (ibid., p. 203).[1] Heiman similarly used multiple machines to measure male and female responses to erotic stimuli as part of her dissertation research, and found that "the more sensitive measure for both sexes was genital pulse amplitude" as opposed to vaginal blood flow (Heiman 1977, p. 272). VPA and penile pulse amplitude could support comparative genital research together, but generally in the 1970s and 1980s, researchers focused on men or women separately.

The vaginal photoplethysmograph has remained the most popular genital measuring device for the last three-and-a-half decades. According to two physicians who have reviewed the current literature on the vaginal photometer, "the VPA (vaginal pulse amplitude) appears to be the most reliable, specific, and sensitive measurement with larger amplitudes reflecting higher levels of blood flow" (Woodward and Diamond 2009, p. 22). However, critics have repeatedly pointed out a variety of problems with the instrument. For example, it may actually be capturing multiple physiological processes, not just sexual arousal. Women

---

[1] A team of British resources tried out a portable device that recorded the movements of a vaginal photoplethysmograph, as well as an electroencephalogram and an electrocardiogram on a portable tape recorder, which the subject then attached to herself with electrodes. Though the subject could make recordings at home in private, the device was awkward and uncomfortable, and apparently the device was not used again. See (Sarrel et al. 1977).

cannot use it during menstruation; it is not adjustable to account for anatomical, moisture, or temperature variation in women's vaginas; it does not provide any anatomical information; and the units of measurement (millivolts) are arbitrary, rather than relative or absolute (Janssen et al. 2007; Heiman and Maravilla 2007). Millivolts are a measurement of the height of a beam of light from the photocell upward into the vagina, which the researcher must calculate from a baseline measurement before any sexual arousal begins. As the baseline measurement varies from person to person, and from session to session due to changes in the menstrual cycle, the baseline for each individual has to be recalculated for every session (Henson et al. 1977). So the statistics and mathematics of the devices are particularly complicated, especially as compared with the absolute measurements that the penile strain gauge provides.

The device has limitations as to what it can measure and when. It cannot isolate its recordings of VPA alone, so it records movement artifacts having nothing to do with sexual arousal. Therefore, the subject must stay relatively still during recording, and it cannot be used for sleep studies (ibid.). Researchers who tried to use the vaginal photoplethysmograph while women masturbated found that the machine recorded large movement artifacts instead of the VBV or VPA (Geer and Quartararo 1976; van Dam et al. 1976; Gillian and Brindley 1979; Henson et al. 1982). Recording during orgasm "as a means of quantifying 'maximum genital arousal,' is not possible because pelvic floor contractions during orgasm create artifacts in the signal, distorting the vasocongestive response" (Chivers et al. 2009, p. 49). Thus the maximum height of a light pulse during vaginal arousal in a single woman, much less in women generally, remains unknown. In other words, a machine specifically designed to investigate women's sexual arousal is unable to register the most marked and significant peak of that arousal. John Bancroft pointed out the challenges of reading vaginal photometer measurements during a conference on the psychophysiology of sex at the Kinsey Institute in Bloomington, Indiana. Researchers do not know "whether, when we're looking at the VPA changing, we're looking at a vascular response pattern that may be distinct from that more specifically, related to tumescence in the clitoris" (Bancroft 2007, p. 58). The key problem with the vaginal photoplethysmograph is what, exactly, it measures at all: "Although both [DC and AC] signals have been found to reflect responses to erotic stimuli, their exact nature and source is unknown" (Janssen et al. 2007, p. 254; see also Hoon et al. 1976; Geer et al. 1974).

Geer, the inventor of the vaginal photoplethysmograph, is well aware of its limitations nearly forty years after its initial appearance in academic literature. He wrote in a 2005 article on the history of the device that "I have been disappointed that a more physiological[ly] appropriate device that is both easy to use and relatively inexpensive has not made an appearance" (Geer 2005, p. 286). Many critiques of the device have appeared repeatedly in the literature over the past thirty years (Henson et al. 1977; Heiman and Maravilla 2007; Janssen et al. 2007; Woodward and Diamond 2009). However, it remains in wide use today due to its inexpensiveness, comfort for the subject, ease of insertion by the subject, ability to take measurements over long periods of time, and ability to do multiple, sequential recordings (Woodward and Diamond 2009). The sex research community has generally, if uneasily, accepted that the vaginal photoplethysmograph has some correlation with women's sexual arousal, even though the precise physiological nature of women's orgasm, as measured with the device, is unknown. Moreover, the authors of a comprehensive 2009 study admitted that "we do not know what constitutes a maximum genital response in women, or whether there is a maximum genital response in women equivalent to a full erection in men" (Chivers et al. 2009, p. 49). Basic physiological questions, such as the nature of peak sexual arousal in women, remain understudied.

Others have kept experimenting regarding different instruments for women's genitals. The invention of a clip that measures labial temperature is another specifically genital measurement for women in which academic interest has waxed and waned following the first published account of its use in 1977 (Henson et al. 1977, 1978; Henson and Rubin 1978). The then-PhD student Donald E. Henson led a group of researchers at the School of Medicine and Rehabilitation Institute, Southern Illinois University (SIU), in the development of the clip. They proceeded on the assumption that surface skin temperature was an excellent indicator of blood volume changes in peripheral tissue, and thus a temperature measure close to the vagina would signify increased blood flow (Henson et al. 1977). The investigation attempted to determine if mental sexual stimulation would produce changes in temperature of the labia minora, which in turn could be used as an indicator of women's overall sexual arousal.

The SIU researchers modified three Yellow Springs Instruments surface-temperature probes so that one attached to the subject's labia, one to her breast, and one to the wall (the wall probe measured ambient temperature). Each of the ten subjects was able to adjust the device herself, and each one watched a "control" film and an erotic film with the device in place, while also reporting her subjective feelings of arousal. Eight of the women's subjective feelings matched their physiological temperature changes, while two of the women's feelings and temperature changes had not. Additionally, three women's labial temperatures did not return to their original baseline, likely because women stayed aroused after the erotic film was over—a situation unmatched in similar male cases. According to the authors, "several subjects reported that it had been exceedingly difficult to refrain from engaging in sexual fantasy during the period following the erotic film presentation. It should be noted that similar reports have never been made by any of the many male subjects who have participated in comparable studies in this laboratory" (ibid., p. 409). The same researchers conducted a different experiment using the labial thermistor and both VPA and VBV measurements with the vaginal photoplethysmograph at once, and found that the baseline measurements returned unevenly across their study populations (Henson and Rubin 1978; Henson et al. 1979b). They asked but did not answer this question: "Why does blood volume subside more for some subjects, pressure pulse more for others, and the labial measure more for still others?" (Henson et al. 1979a, p. 467).

The labial thermistor has some advantages over the vaginal photoplethysmograph: it is not as intrusive as the vaginal photoplethysmograph, it can be used during menstrual periods, it uses an absolute scale of measurement (in degrees), and it does not pick up many movement artifacts. Unfortunately, it requires precise temperature control and does not always return to baseline levels after experimental use (Henson et al. 1979a, b; Woodward and Diamond 2009). The SIU study also showed that aspects of women's sexual physiology were eluding researchers, and suggested the need for specific tools and experimental protocols that captured aspects of women's arousal that tools designed for men or for a nonspecific gender might miss. The study also illustrates two ongoing issues with sex research: that sexual physiology studies sometimes do not capture the full measure of a subject's experiences, if the resolution stage lasts longer than an average testing session; and that the lengths of experimental sessions tend to be based on how well they measure the response cycle of an average male subject, rather than that of an average female subject.

Another key issue in researching women's sexual arousal was and remains how the device connected to the subject. Though the vaginal photoplethysmograph and the labial thermistor were attached to a female subject's genitals, they both required that the subject touch and manipulate her genitals in a way that could feel uncomfortable or invasive. Researchers at two different universities in southern California together looked for measurement devices that did not have to be attached to the subject at all, and decided to record images of men and women masturbating to orgasm in front of a thermographic machine (Abramson et al. 1981). They began "an ambitious research program" using a thermography scanner that was originally designed to detect temperature changes in women with breast cancer (Hoon 1984, p. 770; Abramson et al. 1981). According to the authors, "the temperature 'maps' or thermograms can be qualitatively and quantitatively analyzed, thus making it possible to detect subtle elevations in temperature due to excessive cellular and metabolic activity" (Seeley et al. 1980, pp. 77–78). The researchers used a UTI-Spectrotherm LWIR thermograph machine in their attempt to map the four-stage response cycle that Masters and Johnson outlined (Masters and Johnson 1966).

The thermographic machine recorded multiple images of both subjects, one man and one woman, and the authors included four images of each subject in their article (Seeley et al. 1980). Their findings regarding patterns of heat in the body throughout the response cycle matched those of Masters and Johnson, who observed from the outside patterns, redness, and perspiration on the whole human body throughout the four-stage cycle. The California researchers made unique observations about heat in peripheral parts of the body, such as that the subjects' thumbs were cool, but that their thumbnails were warm during the excitement and orgasm phases of the cycle, "suggesting that this is a highly vascular area" (ibid., p. 83). The man's scrotum stayed the same temperature as his core body temperature, so that the heightened temperature in his penis and other parts of his groin did not affect his sperm. Blood flow increased in the woman subject's pelvic region through the plateau phase until orgasm, and then remained pooled there longer than blood pooled in the man's pelvic region. In other words, the woman's post-orgasmic body was slower to return to a resting state than was the man's body (ibid.). That discovery, which suggested that women's and men's bodies handled the aftereffects of orgasm differently, echoes the labial thermistor researchers' findings that women's resolution phases tend to be longer than men's resolution phases (Henson et al. 1978). That observation has largely remained an intriguing footnote in comparative studies of male and female sexual physiology. It correlates, however, with Roy Levin's observation that "there does not appear to be any obvious explanation for [women] having longer orgasms than men and there are no (as yet) known reasons for the differences in striated muscular activity between men and women's relation to their contractile activity at orgasm" (Levin 2007b, p. 43).[2] The machine that can measure physiological changes in women and men in equal capacity has raised questions about gender similarity and difference in orgasmic capacity that remain unanswered.

Only one other sex research group has continued the use of the thermograph, likely due to its expense and the need for specifically trained technicians for data processing and analysis (Kukkonen et al. 2007). A 2007 Canadian study made clear that the physiological measure of genital temperature corresponded directly to participants' subjective experience of their own sexual arousal. Additionally, the thermograph's gender-neutrality, noninvasive nature, absolute scale of measurement, the provision of full-body data to study as a temperature map and statistically, and the ability to use it during women's menstruation, all mark it as a promising instrument for future studies (Abramson et al. 1981; Kukkonen et al. 2007; Woodward and Diamond 2009).

As of the mid-1980s, researchers had tested and had created a variety of machines, however imperfect, focused on or used for sex research on women. By that time as well, sex researchers had accepted and tested the validity of machines used specifically for men, such as the mercury-in-rubber penile strain gauge that John Bancroft and colleagues used with homosexual men and David H. Barlow and colleagues used with homosexual and heterosexual men (Bancroft et al. 1966; Barlow et al. 1970). The penile strain gauge, which for a time was used as part of aversion therapy, became disassociated with aversion therapy and became—with some modifications, such as replacing the mercury with indium-gallium alloy—a standardized tool in sex research laboratories across the Western world (Heiman 1977; Farkas et al. 1979; Earls and Jackson 1981; Richards et al. 1985; Janssen et al. 2007). An electromechanical penile strain gauge, which is slightly more sensitive during early stages of erection than the mercury or indium-gallium strain gauge, was also developed in the early 1980s (Janssen et al. 1997, 2007). Volumetric penile measurement devices that were originally used for the detection of homosexuals in the Czechoslovakian army or homosexual aversion therapy also became part of sex researchers' standard toolkit, though state and local governments in the United States also presently use them with sex offenders (Freund 1957, 1963, 1965;

---

[2] Levin also made an extended critique of VPA as a measurement tool at a 2003 conference. See (Levin 2007a).

Freund et al. 1958; Abel et al. 1998; Seto 2007). The devices themselves did not change, but the purposes to which researchers have put them have.

The newest instrument devised for measuring changes in men's genitals is the RigiScan, which has its own advantages and disadvantages. The RigiScan, which takes continuous recordings of penile rigidity and tumescence, was introduced to urologists in 1985 for erectile dysfunction, but sex researchers began using it as a more sensitive and complete measurement of men's sexual arousal than the penile strain gauge (Bradley et al. 1985; Munoz et al. 1993). Researchers are able to measure a penis's buckling force and axial rigidity in addition to its circumference. However, it is expensive ($10,000, plus $240 for a box of twelve loop covers), it has not been validated to the satisfaction of sex physiologists, and they are unsure how to apply the buckling force and axial rigidity measurements to specific sexual problems that their subjects face (Sipski et al. 2007). As of spring 2013, the company that manufactured the devices, Timm Medical, has discontinued the RigiScan (Janssen 2013). According to Erick Janssen, who has used the device in his own work, "When it comes to rigidity as an 'endpoint,' we do not know how hard is hard enough....So nowadays 60 %, in RigiScan land, is hard enough. It indicates that we are not talking about absolute truths, but predictions" (ibid., p. 434). Scientists using the RigiScan face a similar dilemma as those working with the vaginal photoplethysmograph: how to contextualize machine-gathered data that they do not wholly understand.

Conclusion

Human Sexual Response and second-wave feminism were both influences in the academic turn toward creating machines oriented toward women's physiology (Masters and Johnson 1966). Vaginal measurement instruments were developed in the 1970s, when the first generation of female researchers in graduate school for psychology and then holding doctorates in the subject brought ideas and ideals of women's empowerment, rights, and agency to their work. The vaginal photometer as an object encapsulated a new feminist sensitivity toward designing sexual physiology experiments that were more in concert with women's anatomy than previous experiments were. With the application of some artificial lubrication, the device fit easily inside a woman's vagina with little discomfort. The subject could insert the device herself, with minimal instruction or intrusion from the researcher. As researchers chose the sexual stimuli they used, whether audio, video, or written, they made sure to include materials depicting sexual activity from a woman's perspective (usually heterosexual, but often also homosexual). Research projects with women as lead investigators, or who had women in positions of influence on research teams, took the lead in developing research devices that made research more comfortable and less invasive for female study subjects.

Machines continue to provide new views of men's and women's bodies as sites of actual and potential sexual desire and arousal, anatomically and physiologically (Geer 1975). Researchers have now developed, modified, and discarded, dozens of different machines to view or to measure aspects of the interiors of women's bodies, through the vaginal photometer, a labial thermistor, or with imaging machines such as MRIs or ultrasounds (Heiman and Maravilla 2007; Everaerd et al. 2007). Machine-created data capturing images of the sexual spaces of women's bodies have become a permanent part of the research conversation surrounding women's sexuality. However, as one survey of machine-gathered sex research put it, "the nature of the female sexual response is highly complex, and is influenced by many factors that incorporate psychosocial as well as physiologic entities....There are a variety of tools that exist to measure sexual response within both of these realms, however many of these tools are limited by their invasiveness, lack of correlation with subjective response, and lack of validation" (Woodward and Diamond 2009, p. 34). There are now multiple machines available that measure aspects of both women's and men's sexuality, even though researchers find serious flaws with a number of them. The subtitle of a 1984 survey article called sex research on women "an unfulfilled promise," and many possibilities for research on sexual physiology are still unfulfilled (Hoon 1984, p. 767). The lack of a machine to measure sexual arousal in women that satisfies psychologists and physiologists alike is itself an unfulfilled promise of sexual science to its practitioners, its readership, and the many people whom such research could affect.

The establishment of standardized machines for research in the field of women's sexuality opens up many new avenues for investigation. It also highlights the current theoretical interest in physiological measurements as a means of studying gender similarity or difference. Sex researchers in the last three decades have usually compared physiological measurements of subjects' sexual arousal with subjective measurements of arousal using a questionnaire or interview, and many have found that men's correlations of physiological and subjective arousal are much higher than women's correlations (Wincze et al. 1977; Hatch 1979). A recent meta-analysis of 130 studies comparing arousal patterns in men and women found that women correlated their physiological and subjective arousals 29 percent of the time, while men correlated theirs 66 percent of the time (Chivers et al. 2009, p. 36). While such a discrepancy led some researchers to consider evolutionary explanations for it, the difference may also be related to the difficulty of comparing women's and men's physiologies when the vaginal photoplethysmograph's and the penile strain gauge's readings can be difficult to interpret, let alone compare: "The gender difference in concordance...might be the result of methodological factors" (Chivers et al. 2009, p. 6). A recent Canadian study using the thermograph pointed that out as well, stating that "the lack of parallel measurement with the most widely used instruments leaves open the possibility that reported differences in sexual arousal between men and women may be the result of technological rather than actual gender differences" (Kukkonen et al. 2007, p. 94).

Furthermore, it is also possible that "the reported lower correlations between physiological and subjective measures of sexual arousal in women as opposed to men may be the result of instrumentation artifacts" (ibid., p. 94). Sex research of all sorts, and particularly sex research on women, is subject to pop psychology interpretations of its meaning and import on women's lives and behaviors, and researchers must be particularly careful in their interpretation of their findings. For example, VBV, VPA, and the labial thermistor all correlate with each other with sexual stimuli but not with other kinds of stimuli, such as anger, fear, or anxiety (Chivers et al. 2009). Regardless of whether researchers focused on VBV, VPA, or the labial thermistor, affective responses and physiological measures do not always match, suggesting that different components of sexual arousal are under the control of different bodily mechanisms (ibid.). However, that discrepancy may be due to the available machines rather than any innate differences between men and women. The discrepancy may also obscure the greater differences that exist within rather than between gendered groups, and it highlights the complicated nature of using gender as a category of analysis when scientists seek causal effects for the responses and behaviors that they study.

Parity between the number of machines for sex research for men and women has largely been achieved over the past four decades. The concurrent use of gender-specific machines and gender-neutral machines (such as the blood pressure cuff and thermograph) shows that the use of both gender-neutral and gender-specific machines together can forward a broad research agenda toward mapping and comprehending the myriad complexities of human sexuality. Likewise, comparative studies that show women having a longer post-orgasmic resolution period than men—a resolution period that scientists noticed but left unmeasured—illustrates the need for serious, ongoing consideration of experimental protocols that capture the entirety of an individual's sexual experience in a laboratory setting (Henson et al. 1977; Seeley et al. 1980). The availability of both kinds of machines provides fertile ground for researchers to consider larger issues of sexuality and gender, and the definitions and meanings of gendered and sexualized categorization. Machine-made findings in sex research raise basic questions about how human sexualities can be similar or different based on gender, identity, erotic preference, and many other factors.

References

Abel, G.G., et al. 1998. Visual reaction time and plethysmography as measures of sexual interest in child molesters. *Sexual Abuse: A Journal of Research and Treatment* 10(2): 81–95. doi:10.1177/107906329801000202.

Abramson, P.R., et al. 1981. Thermographic measurement of sexual arousal: A discriminant validity analysis. *Archives of Sexual Behavior* 10(2): 171–176. doi:10.1007/BF01542177.

Bancroft, J. 2007. Discussion paper. In *The psychophysiology of sex*, ed. E. Janssen, 57–60. Bloomington: Indiana University Press.

Bancroft, J.H.J., H.G. Jones, and B.R. Pullan. 1966. A simple transducer for measuring penile erection, with comments on its use in the treatment of sexual disorders. *Behaviour Research and Therapy* 4(3): 239–242. doi:10.1016/0005-7967(66)90075-1.

Bardwick, J.M., and S.J. Behrman. 1967. Investigation into the effects of anxiety, sexual arousal, and menstrual cycle phase on uterine contractions. *Psychosomatic Medicine* 29(5): 468–482.

Barlow, D.H., et al. 1970. A mechanical strain gauge for recording penile circumference change. *Journal of Applied Behavior Analysis* 3(1): 73–76. doi:10.1901/jaba.1970.3-73.

Bentler, P.M., and P.R. Abramson. 1981. The science of sex research: Some methodological considerations. *Archives of Sexual Behavior* 10(3): 225–251. doi:10.1007/BF01543077.

Bercovici, B., and Y. Palti. 1967. Comparison between uterine, upper vaginal, lower vaginal, and digital pulse. *American Journal of Obstetrics and Gynecology* 98(3): 414–418.

Bercovici, B., G. Uretzki, and Y. Palti. 1972. The effects of estrogens on cytology and vascularization of the vaginal epithelium in climacteric women. *American Journal of Obstetrics and Gynecology* 113(1): 98–103.

Bercovici, B., A. Schechter, and J. Golan. 1973. Cyctolysis in normal and complicated pregnancy. *American Journal of Obstetrics and Gynecology* 116(6): 831–834.

Bergner, D. 2009. What do women want? http://www.nytimes.com/2009/01/25/magazine/25desire-t.html?_r=1&. Accessed 13 Apr 2013.

Berry, M.F. 1988. *Why ERA failed: Politics, women's rights, and the amending process of the constitution.* Bloomington: Indiana University Press.

Bradley, W.E., et al. 1985. New method for continuous measurement of nocturnal penile tumescence and rigidity. *Urology* 26(1): 4–9. doi:10.1016/0090-4295(85)90243-2.

Bullough, V.L. 1994. *Science in the bedroom: A history of sex research.* New York: Basic Books.

Chivers, M., et al. 2009. Agreement of self-reported and genital measures of sexual arousal in men and women: A meta-analysis. *Archives of Sexual Behavior* 39(1): 5–56. doi:10.1007/s10508-009-9556-9.

Cohen, H.D., and A. Shapiro. 1971. A method for measuring sexual arousal in the female. *Psychophysiology* 8(2): 251–252. doi:10.1111/j.1469-8986.1971.tb00456.x.

Cohen, H.D., R.C. Rosen, and L. Goldstein. 1976. Electroencephalographic laterality changes during human sexual orgasm. *Archives of Sexual Behavior* 5(3): 189–199. doi:10.1007/BF01541370.

Earls, C.M., and D.R. Jackson. 1981. The effects of temperature on the mercury-in-rubber strain gauge. *Behavioral Assessment* 3: 145–149.

Everaerd, W., et al. 2007. General discussion. In *The psychophysiology of sex*, ed. E. Janssen, 61–65. Bloomington: Indiana University Press.

Farkas, G.M., L.F. Sine, and I.M. Evans. 1979. The effects of distraction, performance demand, stimulus explicitness and personality on objective and subjective measures of male sexual arousal. *Behaviour Research and Therapy* 17(1): 25–32. doi:10.1016/0005-7967(79)90047-0.

Fisher, S., and H. Osofsky. 1967. Sexual responsiveness in women: Psychological correlates. *Archives of General Psychiatry* 17(2): 214–226. doi:10.1001/archpsyc.1967.01730260086013.

Fisher, S., and H. Osofsky. 1968. Sexual responsiveness in women: Physiological correlates. *Psychological Reports* 22(1): 215–226. doi:10.2466/pr0.1968.22.1.215.

Freund, K. 1957. Diagnostika homosexuality u muza [Diagnosing homosexuality in men]. *Cs Psychiatry* 53: 382–393.

Freund, K. 1963. A laboratory method for diagnosing predominance of homo- or hetero-erotic interest in the male. *Behaviour Research and Therapy* 1(5): 85–93. doi:10.1016/0005-7967(63)90012-3.

Freund, K. 1965. Diagnosing heterosexual pedophilia by means of a test for sexual interest. *Behaviour Research and Therapy* 3: 229–234. doi:10.1016/0005-7967(65)90031-8.

Freund, K., J. Diamant, and V. Pinkava. 1958. On the validity and reliability of the phalloplethysmographic (Php) diagnosis of some sexual deviations. *Review of Czechoslovak Medicine* 4: 145–151.

Geer, J.H. 1975. Direct measurements of genital responding. *American Psychologist* 30(3): 415–418. doi:10.1037/0003-066X.30.3.415.

Geer, J.H. 2005. Development of the vaginal photoplethysmograph. *International Journal of Impotence Research* 17(3): 285–287. doi:10.1038/sj.ijir.3901204.

Geer, J.H., and J.D. Quartararo. 1976. Vaginal blood volume responses during masturbation. *Archives of Sexual Behavior* 5(5): 403–413. doi:10.1007/BF01541333.

Geer, J.H., P. Morokoff, and P. Greenwood. 1974. Sexual arousal in women: The development of a measurement device for vaginal blood volume. *Archives of Sexual Behavior* 3(6): 559–563. doi:10.1007/BF01541137.

Gerhard, J.F. 2001. *Desiring revolution: Second-wave feminism and the rewriting of American sexual thought, 1920 to 1982*. New York: Columbia University Press.

Gillian, P., and G.S. Brindley. 1979. Vaginal and pelvic floor responses to sexual stimulation. *Psychophysiology* 16(5): 471–481. doi:10.1111/j.1469-8986.1979.tb01507.x.

Gordon, L. 2002. *The moral property of women: A history of birth control politics in America*. Urbana: University of Illinois Press.

Green, R. 1971. A statement of purpose. *Archives of Sexual Behavior* 1(1):i. doi:10.1007/BF01540932.

Green, R. 1985. The international academy of sex research: In the beginning. *Archives of Sexual Behavior* 14(4): 293–302. doi:10.1007/BF01550845.

Hatch, J.P. 1979. Vaginal photoplethysmography: Methodological considerations. *Archives of Sexual Behavior* 8(4): 357–374. doi:10.1007/BF01541879.

Heiman, J.R. 1977. A psychophysiological exploration of sexual arousal patterns in females and males. *Psychophysiology* 14(3): 266–274. doi:10.1111/j.1469-8986.1977.tb01173.x.

Heiman, J.R. 2013. Personal communication to the author, 5 Mar. In author's possession.

Heiman, J.R., and K.R. Maravilla. 2007. Female sexual arousal response using serial magnetic resonance imaging with initial comparisons to vaginal photoplethysmography. In *The psychophysiology of sex*, ed. E. Janssen, 103–128. Bloomington: Indiana University Press.

Henson, D.E., and H.B. Rubin. 1978. A comparison of two objective measures of sexual arousal in women. *Behaviour Research and Therapy* 16(3): 143–151. doi:10.1016/0005-7967(78)90060-8.

Henson, D.E., et al. 1977. Temperature change of the labia minora as an objective measure of female eroticism. *Journal of Behavior Therapy and Experimental Psychiatry* 8(4): 401–410. doi:10.1016/0005-7916(77)90011-8.

Henson, D.E., H.B. Rubin, and C. Henson. 1978. Consistency of the labial temperature change measure of female eroticism. *Behaviour Research and Therapy* 16(2): 125–129. doi:10.1016/0005-7967(78)90050-5.

Henson, C., H.B. Rubin, and D.E. Henson. 1979a. Women's sexual arousal concurrently assessed by three genital measures. *Archives of Sexual Behavior* 8(6): 459–469. doi:10.1007/BF01541413.

Henson, D.E., H.B. Rubin, and C. Henson. 1979b. Analysis of the consistency of objective measures of sexual arousal in women. *Journal of Applied Behavior Analysis* 12(4): 701–711. doi:10.1901/jaba.1979.12-701.

Henson, D.E., H.B. Rubin, and C. Henson. 1982. Labial and vaginal blood volume responses to visual and tactile stimuli. *Archives of Sexual Behavior* 11(1): 23–31. doi:10.1007/BF01541363.

Hoon, P.W. 1984. Physiologic assessment of sexual response in women: The unfulfilled promise. *Clinical Obstetrics and Gynecology* 27(3): 767–780.

Hoon, P.W., J.P. Wincze, and E.F. Hoon. 1976. Physiological assessment of sexual arousal in women. *Psychophysiology* 13(3): 196–204. doi:10.1111/j.1469-8986.1976.tb00097.x.

Janssen, E. 2013. Personal communication to the author, 7 May. In author's possession.

Janssen, E., et al. 1997. An in vivo comparison of two circumferential penile strain gauges: Introducing a new calibration method. *Psychophysiology* 34(6): 717–720. doi:10.1111/j.1469-8986.1997.tb02147.x.

Janssen, E., N. Prause, and J.H. Geer. 2007. The sexual response. In *Handbook of psychophysiology*, 3rd ed, ed. J.T. Cacioppo, L.G. Tassinary, and G.G. Bernston, 245–266. New York: Cambridge University Press.

Johnson, J.W. 2005. *Griswold v. Connecticut: Birth control and the constitutional right of privacy*. Lawrence: University of Kansas.

Jovanovic, U.J. 1971. The recording of physiological evidence of genital arousal in human males and females. *Archives of Sexual Behavior* 1(4): 309–320. doi:10.1007/BF01638059.

Kaplan, H.S. 1974. *The new sex therapy: Active treatment of sexual dysfunctions*. New York: Brunner/Mazel.

Karacan, I., A.L., Rosenbloom, and R.L., Williams. 1970. The clitoral erection cycle during sleep. In *Abstracts of annual meeting of the association for the psychophysiological study of sleep*, ed. R. Cartwright, A. Rechtschaffen, and J. Rhodes. *Psychophysiology* 7(2): 338. doi:10.1111/j.1469-8986.1970.tb02238.x.

Kinsey, A.C., et al. 1953. *Sexual behavior in the human female*. Philadelphia: W. B. Saunders.

Kukkonen, T.M., et al. 2007. Thermography as a physiological measure of sexual arousal in both men and women. *The Journal of Sexual Medicine* 4(1): 93–105. doi:10.1111/j.1743-6109.2006.00399.x.

Levin, R.J. 2007a. Discussion paper. In *The psychophysiology of sex*, ed. E. Janssen, 129–135. Bloomington: Indiana University Press.

Levin, R.J. 2007b. The human sexual response—similarities and differences in the anatomy and function of the male and female genitalia. In *The psychophysiology of sex*, ed. E. Janssen, 35–56. Bloomington: Indiana University Press.

Maier, T. 2009. *Masters of sex: William Masters and Virginia Johnson, the couple who taught America how to love*. New York: Basic Books.

Mansbridge, J.L. 1986. *Why we lost the ERA*. Chicago: University of Chicago Press.

Masters, W.H., and V.E. Johnson. 1961a. The physiology of the vaginal reproductive function. *Western Journal of Surgery, Obstetrics, and Gynecology* 69(2): 105–120.

Masters, W.H., and V.E. Johnson. 1961b. The artificial vagina: Anatomic, physiologic, psychosexual function. *Western Journal of Surgery, Obstetrics, and Gynecology* 69(3): 192–212.

Masters, W.H., and V.E. Johnson. 1966. *Human sexual response*. Boston: Little, Brown.

May, E.T. 2010. *America and the pill: A history of promise, peril, and liberation*. New York: Basic Books.

Mims, S.S. 1972. Eisenstadt v. Baird: Massachusetts statute prohibiting distribution of contraceptives to unmarried persons held unconstitutional. *Southwestern Law Journal* 26(4): 775–780.

Morrow, R. 2008. *Sex research and sex therapy: A sociological analysis of Masters and Johnson*. New York: Routledge.

Munoz, M.M., J. Bancroft, and I. Marshall. 1993. The performance of the Rigiscan in the measurement of penile tumescence and rigidity. *International Journal of Impotence Research* 5(2): 69–76.

Ohlmeyer, P., and H. Brilmayer. 1947. Periodische vorgange im schlaf [Periodic events in sleep]. *Pflüger's Archiv für die gesamte Physiologie des Menschen und der Tiere* 248: 559–560. doi:10.1007/BF00362669.

Ohlmeyer, P., H. Brilmayer, and H. Hullstrung. 1944. Periodische vorgange im schlaf II [Periodic events in sleep II]. *Pflüger's Archiv für die gesamte Physiologie des Menschen und der Tiere* 249: 50–55. doi:10.1007/BF01764449.

Palti, Y., and B. Bercovici. 1967. Photoplethysmographic study of the vaginal blood pulse. *American Journal of Obstetrics and Gynecology* 97(2): 143–153.

Richards, J.C., et al. 1985. A controlled investigation into the measurement properties of two circumferential penile strain gauges. *Psychophysiology* 22(5): 568–571. doi:10.1111/j.1469-8986.1985.tb01653.x.

Roach, M. 2008. *Bonk: The curious coupling of sex and science*. New York: W. W. Norton.

Sarrel, P.M., J. Foddy, and J.B. McKinnon. 1977. Investigation of human sexual response using a cassette recorder. *Archives of Sexual Behavior* 6(4): 341–348. doi:10.1007/BF01541205.

Seeley, D.M., et al. 1980. Thermographic measurement of sexual arousal: A methodological note. *Archives of Sexual Behavior* 9(2): 77–85. doi:10.1007/BF01542260.

Seto, M.C. 2007. Psychophysiological assessment of paraphilic sexual interests. In *The psychophysiology of sex*, ed. E. Janssen, 475–491. Bloomington: Indiana University Press.

Shapiro, A., et al. 1968. Vaginal blood flow changes during sleep and sexual arousal. In: *Abstracts of papers presented to the seventh annual meeting of the association for the psychophysiological study of sleep*, ed. D. Foulkes, and A. Kales. *Psychophysiology* 4(3):394. doi:10.1111/j.1469-8986.1968.tb02781.x.

Sintchak, G., and J.H. Geer. 1975. A vaginal plethysmograph system. *Psychophysiology* 12(1): 113–115. doi:10.1111/j.1469-8986.1975.tb03074.x.

Sipski, M., et al. 2007. General discussion. In *The psychophysiology of sex*, ed. E. Janssen, 429–435. Bloomington: Indiana University Press.

Tone, A. 2001. *Devices and desires: A history of contraceptives in America*. New York: Hill and Wang.

Uretzky, G., and Y. Palti. 1977. Elastic properties of blood vessels determined by photoelectric plethysmography. *Angiology* 28(2): 75–83. doi:10.1177/000331977702800202.

van Dam, F.S.A.M., et al. 1976. Sexual arousal measured by photoplethysmography. *Behavioral Engineering* 3: 97–101.

Watkins, E.S. 2001. *On the pill: A social history of contraceptives, 1950–1970*. Baltimore: Johns Hopkins University Press.

Wincze, J.P., P. Hoon, and E.F. Hoon. 1977. Sexual arousal in women: A comparison of cognitive and physiological responses by continuous measurement. *Archives of Sexual Behavior* 6(2): 121–133. doi:10.1007/BF01541704.

Woodward, T.L., and M.P. Diamond. 2009. Physiologic measures of sexual function in women: A review. *Fertility and Sterility* 92(1): 19–34. doi:10.1016/j.fertnstert.2008.04.041.

Zuckerman, M. 1971. Physiological measures of sexual arousal in the human. *Psychological Bulletin* 75(5): 297–329. doi:10.1037/h0030923.

Chapter 5
Conclusion: The Future of Human Sex Research Technologies

Abstract This chapter summarizes the main ideas of this book on machines used in sex research, a history which began in the late nineteenth century in Berlin and continued through the late twentieth century in the United Kingdom and the United States. It outlines some new machines developed in the early twenty-first century, including the arousometer and a hands-off clitoral vibrator. Sex research using machines in the present continues to face challenges, including the need for more accurate and specific machines to measure women's sexual arousal, a lack of standardization in practices among subsets of sex researchers, the ongoing popularity of aversion therapy for homosexuality despite a near-total failure rate, and the need to raise funding for basic as well as applied physiological sex research. The history of machines used in sex research illustrates the lack of linearity in scientists' pursuit of high-quality and useful measurements over the past century, and their continued determination to develop a unified theory of human sexuality.

Keywords Sex research · Human sexuality · Vibrator · Sexual arousal · Women's sexuality

The history of machines used in sex research has been one of exploration, invention, failure, frustration, and professional acceptance, all linked by researchers' desire to know, and in some cases, police, the human body as a sexual entity. One of the aims of this book has been to "trace the career of material things as they move through different settings and are attributed value" (Takeshita 2011, p. 4). As the use of different kinds of devices has waxed and waned from the end of the nineteenth through the first two decades of the twenty-first century, sex researchers' intentions for inventing and using them has shifted according to the political climate in which the research has taken place. Some devices, such as the blood pressure cuff, cardiograph, vaginal photoplethysmograph, penile strain gauge, RigiScan, television or monitor, video or DVD player, video or audio recorder, and computer have all become common, indeed essential, parts of physiological sex research. Tracing the creation, disuse, and reinvention of machines used in sex research supports Wiebe E. Bijker, Thomas P. Hughes, and Trevor J. Pinch's multidirectional model of technological development

D. J. Drucker, *The Machines of Sex Research*,   91
SpringerBriefs in History of Science and Technology,
DOI: 10.1007/978-94-007-7064-5_5, © The Author(s) 2014

(Bijker et al. 1987; Pinch and Bijker 1987). Researchers invented and used existing machines to study human sexuality in the medical, political, and cultural contexts specific to their historical time.

Older machines regularly undergo validity testing and improvement, and new machines are continually in development. Some more recent devices include the arousometer, which consists of a computer optical mouse mounted on a wooden track. The track is divided into ten equally spaced intervals, and participants move the mouse up and down the track with their hand to indicate increases or decreases in their arousal (Rellini et al. 2005; Roach 2008). Another new machine is a hands-off clitoral vibrator that provides "vibrotactile" stimulation, used for experiments in which participants watch erotic videos with and without clitoral stimulation from it (Ponseti and Bosinski 2010). Sex researchers' interest in discovery inspires mechanical creativity, which sometimes harms their subjects, sometimes helps them, and sometimes leaves them right where they started.

Despite the efforts of many involved in liberation movements at all levels of society in the past half century, research in the human sciences designed for punishing or deliberately inflicting pain continues in the present. Electroconvulsive therapy is still used today worldwide, though not without ethical debate and controversy, largely for schizophrenia, bipolar disorder, severe depression, and related mental disorders (Kellner et al. 2012; Pompili et al. 2013). Psychotheraputic, non-machine-based aversion therapies continue to be popular among a small number of British therapists and conservative Christian religious groups, who encourage individuals torn between their religious faith and their same-sex desires and behaviors to "pray away the gay" (Vinocur 2009; Narth 2013). Such therapies, as Douglas Haldeman has detailed, are themselves plagued with scandal and have a near-complete long-term failure rate (Haldeman 1994, 1999). Furthermore, the major American psychological and psychiatric associations have denounced aversion therapy as contrary to the goals of psychiatry (Anton 2010). So the possibility of aversion therapy for homosexuality remains in the present.

The tension between sex research as a disciplinary instrument versus sex research as a means of expanding knowledge about the human body has never fully disappeared. While American, Canadian, and European sex researchers in the present tends to accept and to support diversity in human sexual behavior, federal and state mandates for abstinence-only education in schools—not to mention recent legislation making aversion therapy for sexual orientation illegal for teenagers in California—shows that some public support remains for research that curtails sexual exploration, understanding, and fulfillment (California State Senate 2012; Eckholm 2012). Progressive sex researchers and their supporters must continue to advocate for research that embraces sexual health for people of all genders, orientations, and preferences. And better research will continue to require better machines. As Vern Bullough wrote at the end of his survey of American and European sex research, "there are still a lot of things we need to know and much more that we would like to know, if enough money, time, and professionals can be found" (Bullough 1994, p. 299).

Scientists have even surprised themselves as they have explored the possibilities—whether forward-thinking or harmful—of machines that they or their colleagues created. Hans-Jörg Rheinberger put it thus: "Scientific objects come into existence as a result of unprecedented events that time and again subvert the finite capacities of imagination of a scientist who remains always embedded in a particular thinking frame and a local experimental culture. They remain objects of research as long as they have the power to manifest themselves in yet unthought-of ways in the future" (Rheinberger 2000, p. 273). Sometimes machines, such as the electric shock box, were used for harm. Some of them, such as the penis camera, were used for basic physiological research related to sexual arousal and response (Masters and Johnson 1966). Some others, such as the clitorophotoplethysmograph, were used only once or twice and then were never mentioned again in the research literature (Zuckerman 1971). Yet others, like the penile strain gauge, were used to cause pain in their early history but later for basic research. Rheinberger argued that scientific objects have multiple possibilities embedded in them as long as they produce new scientific knowledge (Rheinberger 2000). When they fail to produce new knowledge, they disappear from use. Such was and is the case with instruments produced to improve understanding human sexuality.

Scientists of sexuality need to use machines to measure bodily processes and functions. But machines can often obscure as much as they reveal. Nothing in nature tells a human observer how it should be measured; nature is not written in the mathematics that humans use (Boumans 2013). Human judgment combined with mechanical processes together make up any measuring system. Laboratory researchers looked for, and continue to look for, a way to map "the real" and "the natural" of sexual physiology using mechanical tools in artificial environments very far from the average circumstances of most peoples' sexual experiences. The German researchers Hellmuth Klumbies and Gerhard Kleinsorge wrote of their multi-machine laboratory experiments with subjects having some difficulty fantasizing to orgasm, "After all, the environment dampened the senses" (Klumbies and Kleinsorge 1950, p. 958).[1] As one early user of the vaginal photoplethysmograph stated, "the sexual behavior observed in the laboratory environment may thus be only crudely analogous to sex which occurs in the natural setting" (Hatch 1979, p. 371). Since precisely replicating natural or real circumstances under research conditions can be difficult, research findings and results show "the real" of human sexuality, but through a lens. As Bruno Latour and Steve Woolgar wrote in their classic study of laboratory science, "scientific activity is not 'about nature,' it is a fierce fight to construct reality. The laboratory is the work place and the set of productive forces, which makes construction possible" (Latour and Woolgar 1986, p. 243). Some constructions of reality, such as the idea that homosexuality was an illness that physicians could fix through electroshock therapy, have been discredited in reputable scientific and medical communities. Other constructions of

---

[1] "Immerhin wirkte das Milieu im Sinne der Dämpfung" (Klumbies and Kleinsorge 1950, p. 958).

reality, such as the four-stage sexual response cycle that William Masters and Virginia Johnson articulated, remain a critiqued yet active model for research many years after their first introduction into sexual science literature (Masters and Johnson 1966; Tiefer 1995).

As Anne Fausto-Sterling articulated in her discussion of brain science, "simplifying body parts in order to layer some conceptual order onto the daunting complexity of the living body is the daily bread of the working scientist" (Fausto-Sterling 2000, p. 127). Sexual scientists, like human scientists in general, face "the daunting complexity of the living body" every day in their efforts to understand their subject better, using some instruments that are broadly acceptable in their research community but still inadequate. The contemporary sex research community faces a number of challenges and opportunities, including a lack of instruments that measure aspects of the body that they want to measure. Chief among those challenges is maintaining a steady level of funding for sex research, particularly for women's sexuality and for studies of high-risk populations. As HIV and AIDS first appeared in the United States in 1979 and were first acknowledged in the academic community in 1981, many sex researchers turned their attention away from basic physiological research toward understanding those new illnesses. In the early 1980s, "the AIDS epidemic engulfed the attention of the field, putting a priority on prevention and making desire not an emotion to explore but an element to be feared" (Bergner 2009). However, some academics continued their focus on basic sexual physiology, which in turn supported HIV/AIDS research.

In addition, protocols for experimental structure, laboratory setup, data gathering, and data analysis from those instruments are not wholly standardized, which hampers the ability of scientists to compare studies. As Erick Janssen wrote in a recent survey of sexual psychophysiology literature, "In contrast to many other areas of psychophysiological research, no guidelines exist for the measurement of sexual arousal, and the current lack of standardization of signal recording, processing, and analysis, complicates the evaluation and comparison of research findings" (Janssen et al. 2007, p. 258). Many sex researchers, including now former Kinsey Institute director John Bancroft, encourage their peers to use multiple machines in any investigation of physiology to better understand the connections between different bodily systems. Looking at VPA alone is not enough; multiple machines are necessary to take a full measure of the female body. "In our pursuit of VPA...we're going to measure it, with whatever device, we should be looking closely at its relationship to other indices of arousal, such as blood pressure change, pulse pressure change, peripheral vasomotor tone, and so on" (Bancroft 2007, p. 58). Sex researchers continue to conduct basic physiological research using the most appropriate and accurate techniques and machines that are available to them, and they also do applied research in pursuit of solving clinical problems such as sexual arousal and mood disorders. For example, while Viagra, Cialis, and other sildenafil drugs are widely used for male erectile dysfunction, scientists and drug companies alike continue to search for a drug that will stimulate women's sexual arousal (Loe 2004; McLaren 2007; Roach 2008).

Whatever new tools are developed for human sex research in the coming years, researchers will use them in combination with others to get as full a picture as possible of the interactions between body and mind that make up sexual identity, behavior, and experience. Bancroft mused in an article reviewing the legacy of Alfred Kinsey, "there is general agreement in the field of sex research that we understand little about what we refer to as sexual arousal, beyond it being a complex interaction between psychological and physiological processes" (Bancroft 2004, p. 29). The current Kinsey Institute director, Julia R. Heiman, said in a 2009 interview that "No one right now has a unifying theory" of human sexuality (Bergner 2009). As her interviewer wrote, "the interest has brought scattered sightlines, glimpses from all sorts of angles" (ibid.). The synthetic theory of human sexuality that Kinsey envisioned writing in the 1940s and 1950s, with a multiple volume series based on one hundred thousand interviews, is still far in the future, even after sixty additional years of accumulated knowledge (Kinsey et al. 1948, 1953).

Melding histories of technology with histories of gender, sexuality, and feminism, gender and sexuality theory, and spatial theory shows that those histories and theories are inextricably interlinked. While the machinery of sex research for men was designed and implemented in a period of restrictions on male bodies, sexual behavior, and identity, the machinery of sex research for women was created in a period of advocacy for and implementation of new freedoms for sexual and gendered personhood. Historical developments in homosexual and women's rights inform the technologies that shape modern understandings of the anatomy, physiology, gender, and sexuality of the human person. Rheinberger has argued that "nowhere is the realm of the scientific real a closed space" (Rheinberger 2000, p. 274). Human sexuality as an academic field remains an open space, open to countless possible explorations of the scientific real.

References

Anton, B.S. 2010. Proceedings of the American Psychological Association for the legislative year 2009: Minutes of the annual meeting of the Council of Representatives and minutes of the meetings of the Board of Directors. *American Psychologist* 65(5): 385–475. doi:10.1037/a0019553.

Bancroft, J. 2004. Alfred C. Kinsey and the politics of sex research. *Annual Review of Sex Research* 15(1): 1–40. doi:10.1080/10532528.2004.10559818.

Bancroft, J. 2007. Discussion paper. In *The psychophysiology of sex*, ed. E. Janssen, 57–60. Bloomington: Indiana University Press.

Bergner, D. 2009. What do women want? http://www.nytimes.com/2009/01/25/magazine/25desire-t.html?_r=1&. Accessed 13 Apr 2013.

Bijker, W.E., T.P. Hughes, and T.J. Pinch. 1987. Introduction. In *The social construction of technological systems: New directions in the sociology and history of technology*, ed. W.E. Bijker, T.P. Hughes, and T.J. Pinch, 9–15. Cambridge: MIT Press.

Boumans, M. 2013. Clinical measurement. Unpublished conference presentation, Dimensions of Measurement Workshop, Zentrum für Interdisziplinäre Forschung. Germany: Universität Bielefeld.

Bullough, V.L. 1994. *Science in the bedroom: A history of sex research*. New York: Basic Books.

California State Senate. 2012. SB-1172: Sexual orientation change efforts. http://leginfo.legislature.ca.gov/faces/billNavClient.xhtml?bill_id=201120120SB1172. Accessed 11 Apr 2013.

Eckholm, E. 2012. California is first state to ban gay "cure" for minors. New York Times. http://www.nytimes.com/2012/10/01/us/california-bans-therapies-to-cure-gay-minors.html?_r=0. Accessed 11 Apr 2013.

Fausto-Sterling, A. 2000. *Sexing the body: Gender politics and the construction of sexuality*. New York: Basic Books.

Haldeman, D. 1994. The practice and ethics of sexual orientation conversion therapy. *Journal of Counseling and Clinical Psychology* 62(2): 221–227. doi:10.1037/0022-006X.62.2.221.

Haldeman, D. 1999. The pseudo-science of sexual orientation conversion therapy: Clinical and social implications. *Angles: The Policy Journal of the Institute for Lesbian and Gay Strategic Studies* 4(1):1–4. http://www.drdoughaldeman.com/doc/Pseudo-Science.pdf. Accessed 11 April 2013.

Hatch, J.P. 1979. Vaginal photoplethysmography: Methodological considerations. *Archives of Sexual Behavior* 8(4): 357–374. doi:10.1007/BF01541879.

Janssen, E., N. Prause, and J.H. Geer. 2007. The sexual response. In *Handbook of psychophysiology*, 3rd ed, ed. J.T. Cacioppo, L.G. Tassinary, and G.G. Bernston, 245–266. New York: Cambridge University Press.

Kellner, C.H., et al. 2012. ECT in treatment-resistant depression. *American Journal of Psychiatry* 169(12): 1238–1244. doi:10.1176/appi.ajp.2012.12050648.

Kinsey, A.C., W.B. Pomeroy, and C.E. Martin. 1948. *Sexual behavior in the human male*. Philadelphia: W. B. Saunders.

Kinsey, A.C., et al. 1953. *Sexual behavior in the human female*. Philadelphia: W. B. Saunders.

Klumbies, G., and H. Kleinsorge. 1950. Das herz in orgasmus [The heart in orgasm]. *Medizinische Klinik* 45(31): 952–958.

Latour, B., and S. Woolgar. 1986. *Laboratory life: The construction of scientific facts*. Princeton, New Jersey: Princeton University Press.

Loe, M. 2004. *The rise of Viagra: How the little blue pill changed sex in America*. New York: New York University Press.

Masters, W.H., and V.E. Johnson. 1966. *Human sexual response*. Boston: Little, Brown.

McLaren, A. 2007. *Impotence: A cultural history*. Chicago: University of Chicago Press.

Narth. 2013. National Association for Research and Therapy of Homosexuality. http://narth.com. Accessed 24 Apr 2013.

Pinch, T.J., and W.E. Bijker. 1987. The social construction of facts and artifacts: Or how the sociology of science and the sociology of technology might benefit each other. In *The social construction of technological systems: New directions in the sociology and history of technology*, ed. W.E. Bijker, T.P. Hughes, and T.J. Pinch, 17–50. Cambridge: MIT Press.

Pompili, M., et al. 2013. Indications for electroconvulsive treatment in schizophrenia: A systematic review. *Schizophrenia Research* 146(1–3): 1–9. doi:10.1016/j.schres.2013.02.005.

Ponseti, J., and H.A.G. Bosinski. 2010. Subliminal sexual stimuli facilitate genital response in women. *Archives of Sexual Behavior* 39(5): 1073–1079. doi:10.1007/s10508-009-9587-2.

Rellini, A.H., et al. 2005. The relationship between women's subjective and physiological sexual arousal. *Psychophysiology* 42(1): 116–124. doi:10.1111/j.1469-8986.2005.00259.x.

Rheinberger, H. 2000. Cytoplasmic particles: The trajectory of a scientific object. In *Biographies of scientific objects*, ed. L. Daston, 270–294. Chicago: University of Chicago Press.

Roach, M. 2008. *Bonk: The curious coupling of sex and science*. New York: W. W. Norton.

Takeshita, C. 2011. *The global politics of the IUD: How science constructs contraceptive users and women's bodies*. Cambridge: MIT Press.

Tiefer, L. 1995. *Sex is not a natural act and other essays*. Boulder, Colorado: Westview Press.

Vinocur, N. 2009. "Gay cure" therapies still used by few in UK: Study. http://www.reuters.com/article/2009/03/26/us-homosexuality-therapy-idUSTRE52P01120090326. Accessed 24 Apr 2013.

Zuckerman, M. 1971. Physiological measures of sexual arousal in the human. *Psychological Bulletin* 75(5): 297–329. doi:10.1037/h0030923.

Erratum to: The Penile Strain Gauge and Aversion Therapy: Measuring and Fixing the Sexual Body

Erratum to:
Chapter 2 in: D. J. Drucker, *The Machines of Sex Research*,
DOI 10.1007/978-94-007-7064-5_2

In printed version, the citation "(Smith et al. 2004, p. 428)" cited in Chapter 2 was wrong. The correct citation is "(Dickinson 2012, p. 1349)" and the respective reference in list is to be read as Dickinson, T. Cook, M., Playle, J. & Hallett, C. (2012) "Queer" Treatments: giving a voice to former patients who received treatments for their "sexual deviations". Journal of Clinical Nursing, 21 (9), 1345–54.

The online version of the original chapter can be found under
DOI 10.1007/978-94-007-7064-5_2

D. J. Drucker (✉)
Technische Universität Darmstadt, Caroline-Herschel-Str. 14 64293 Darmstadt, Germany
e-mail: drucker@ifs.tu-darmstadt.de

D. J. Drucker, *The Machines of Sex Research*,   E1
SpringerBriefs in History of Science and Technology,
DOI: 10.1007/978-94-007-7064-5_6, © The Author(s) 2014